SPATIAL AUTOCORRELATION
A Primer

Daniel A. Griffith
Department of Geography
State University of New York
at Buffalo

Library of Congress Card Number 87-1180
ISBN 0-89291-192-2

Library of Congress Cataloging in Publication Data

Griffith Daniel A.
Spatial autocorrelation.

(Resource publications in geography)
Bibliography: p.
Includes index.
Spatial analysis (Statistics) 2. Geography —
Statistical methods. I. Title. II. Series.
QA278.2.G75 1987 910'.01'5195 87-1180
ISBN 0-89291-197-2

Publication Supported by the A.A.G.

Graphic Design by CGK

Printed by Commercial Printing Inc.
State College, Pennsylvania

Foreword

It is axiomatic to geography that events in space are not random. When geographers examine the spatial distribution of phenomena, they attempt to understand and explain patterns as resulting from processes, causal linkages between events in time and space. Causal inference is frequently developed through statistical insight which focuses on systematic relationships between phenomena. If events in space are related, and if the distribution of these events can be statistically linked, then statistical inference can be an exceedingly powerful tool to suggest the processes underlying the spatial distribution of events. Beneath obvious statistical analyses such as correlation and regression, however, there lurks a significant pitfall: spatial autocorrelation.

Spatial autocorrelation refers to the essential redundancy of information about a phenomenon which is related to spatial proximity. Sampling units, for example, that are adjacent are more strongly related to each other than to more distant units. Thus in describing the population from which the sample is drawn, or more critically, in using the sample to test statistically for systematic relationships, analysis of and correction for spatial autocorrelation is obligatory. Dan Griffith, in *Spatial Autocorrelation: A Primer,* explains the concept of autocorrelation. Moreover, he rigorously develops the necessary modifications to traditional statistical methods which are necessary to account for this phenomenon.

Resource Publications in Geography are sponsored by the Association of American Geographers, a professional organization whose purpose is to advance studies in geography and to encourage the application of geographic research in education, government, and business. This series traces its origins to the Association's Commission on College Geography, whose *Resource Papers* were launched in 1968. Eventually 28 papers were published under the sponsorship of the Commission through 1974 with the assistance of the National Science Foundation. Continued NSF support after completion of the Commission's work permitted the *Resource Papers for College Geography* to meet the original series goals for an additional four years and sixteen volumes:

> *The Resource Papers have been developed as expository documents for the use of both the student and the instructor. They are experimental in that they are designed to supplement existing texts and to fill a gap between significant research in American geography and readily accessible materials. The papers are concerned with important concepts or topics in modern geography and focus on one of three general themes: geographic theory, policy implications, or contemporary social relevance. They are designed to implement a variety of undergraduate college geography courses at the introductory and advanced level.*

The popularity and usefulness of the two series suggested the importance of their continuation after 1978, once a self-supporting basis for their publication had been established.

For the **Resource Publications,** the original goals remain paramount. However, they have been broadened to include the continuing education of professional geographers as well as communication with the public on contemporary issues of geographic relevance. This monograph was developed, printed, and distributed under the auspices of the Association, whose members served in advisory and review roles during its preparation. The ideas presented, however, are the author's and do not imply AAG endorsement.

Many readers may find the mathematics presented in this volume somewhat challenging. For appropriate derivations, one may refer to the appendices or cited works. Several short microcomputer programs will help to illustrate basic concepts; Griffith also provides the necessary modifications to traditional statistical tests as well as shortcuts to calculation.

This volume is the last in the current Resource Publications series. Although the 1985 series ends with this book, it should be noted that author Griffith's manuscript was received in late 1986 and was initially intended for later publication. Due to earlier projects that failed to materialize, Griffith's work was placed in the annual series of earlier date.

The editor and advisory board are grateful to Griffith for providing this important primer on spatial autocorrelation and commend it to you as you seek further sophistication in your approach to spatial analysis.

C. Gregory Knight, *The Pennsylvania State University*
Editor, Resource Publications in Geography

Preface

One of the distinguishing features of statistical analysis in geography is the phenomenon of spatial autocorrelation, which refers to latent interobservational dependencies. The seminal treatment of this topic in a somewhat comprehensive manner was by Cliff and Ord, who more recently revised their work under the title of *Spatial Processes* (1981). Somewhat similar but less comprehensive treatments can be found in Paelinck and Klaassen's *Spatial Econometrics* (1979), Ripley's *Spatial Statistics* (1981), Ahuja and Schachter's *Pattern Models*, (1983), and Upton and Fingleton's *Spatial Data Analysis by Example* (1985). To date, though, a primer has not been published for this topic, leaving the student of quantitative geography faced with a major jump in complexity of subject matter when moving from introductory books, such as those by Ebdon (1977) and Silk (1979), to one of the more advanced texts. The purpose of this Resource Publication is to fill this void by providing a primer that bridges introductory and advanced texts. The pedagogic approach of this monograph places a strong emphasis on heuristics and intuition, couching most of the materials in statistical terms (especially sampling distribution frameworks) made familiar in such books as those by Ebdon and Silk. Numerous examples that students could contrive are presented in the form of MINITAB computer code. MINITAB is the registered trademark of MINITAB, Inc.

Besides the fact that there is significant need for such a book, another important reason for producing it is that introductory and intermediate quantitative geography courses should benefit from its availability. Rather than being a textbook though, this research monograph will serve as a useful supplement to existing texts in quantitative geography. It also should serve as a reference for many geography instructors and professional geographers outside the academic community.

Daniel A. Griffith

To
Donald S. Griffith, Sr.
and
Jeanne C. Griffith

Contents

List of Figures

List of Tables

1

Statistical Analysis and Autocorrelation

One of the primary reasons that statistics are collected and studied is to gain information about a population. The information gained allows educated and constrained speculation to be put forward, in that reasonable statements about the population can be made. For instance, if a person wanted to know what the median household income in the United States was in 1985, then a guess about this value could be put forth, usually in the form of an hypothesis. But how reasonable is this guess? If a sample of size zero is taken, the sample yields no information about the parent population of all United States households, and almost any quantity could, arguably, be reasonable. However, if a sample of size $n>0$ is selected, then an empirical benchmark exists against which to compare the hypothesized value and assess its reasonableness. Therefore the sample furnishes useful information, information which increases as the sample size increases, such that the sample's information content is of interest to researchers.

The quality of information also is of concern. In this context quality refers to the grade or degree of excellence of the information; whether or not it is ambiguous, accurate, distinct, representative, and the like. One important factor influencing information quality is the manner in which the corresponding sample was drawn. Overall the best information, aside from an examination of the entire population, is obtained from a random sample. Other sampling schemes, such as stratified or systematic sampling, have been devised to enhance particular properties of selected samples, but often at the expense of other sample properties, such as bias and minimum standard errors. But these modified sampling schemes almost always are coupled with random sampling (*e.g.*, stratified random sampling) to maintain quality information. The chief reason that random sampling yields quality information is because the selected observations are independent, and hence furnish incremental amounts of information.

Unfortunately, situations exist in practice in which the quantity and quality of sample information is less than would be desirable. An illustration can serve both to exemplify this point and to set the stage for a discussion of the importance of spatial autocorrelation. Consider a population consisting of three elements, represented here by the set {A,B,C}, each having a corresponding numerical value, represented here by the set {2,4,6}. A company is hired to survey this population and make some statement about the arithmetic average of the single numerical variable. The research firm soliciting the sample feels that a sample size of 5, drawn with replacement and order being important, will provide the necessary quality and quantity of information for the purpose of its study. The company conducting the survey wishes to minimize its survey costs, so it draws a random sample of size 2. However, it uses a rule that the first element selected will appear in the 'sample' twice, and the second element selected will appear in the 'sample' three times. Hence the survey company will be able to supply a 'sample' of size 5 while incurring little expense. It expects that the research firm will think that a sample of five independent drawings has been obtained.

The problem described above has three different sampling distributions affiliated with it (Table 1.1). From the results reported in this Table 1.1, one can see that the survey company actually selected a sample from only a subset of the possibilities envisioned by the research firm. Now the population mean is $\mu = 4$. In addition the sampling distribution means are $\mu_{\bar{x}}(n=2) = 4$, $\mu_{\bar{x}}(\text{weighted } n=5) = 4$, and $\mu_{\bar{x}}(n=5) = 4$. So the survey company did not bias results pertaining to estimation of the arithmetic mean. Meanwhile, $\sigma^2 = 8/3$. Furthermore:

$$\sigma^2_{\bar{x}}(n=2) = 4/3$$
$$\sigma^2_{\bar{x}}(\text{weighted } n=2) = [2^2(8/3) + 3^2(8/3)]/5^2 = 4.16/3 \quad \text{and}$$
$$\sigma^2(n=5) = 1.6/3.$$

Moreover, the underlying sampling distribution variance is 4/3, the survey company used a model to generate the sample for which the sampling distribution variance is 4.16/3, and the research firm thinks that the sampling distribution variance is 1.6/3. Clearly then the quality and quantity of information has been altered.

This information distortion can be attributed to various factors. Foremost is the fact that if element A is selected, and contains a certain amount of information about the parent population, repeating it twice (versus replacing it and again drawing an element randomly) does not double the amount of information available. In fact, this doubling process introduces redundant information. Second, the weighting model used will exaggerate this distortion.

Scrutinizing the sample composition presented to it, the research firm may feel that the sample is suspect. After all, the probability of selecting three of one element, two of a second element, and none of a third element from the sampled population, if random sampling is conducted with replacement and order is important, is 60/243 (see Table 1.1). In contrast, the probability of a sample composition where each element appears at least once is 150/243. So in concluding that something is wrong, the research firm can choose one of two courses of action to follow. First, it could seek to uncover the underlying sample scheme for $n = 2$ that actually was used by the survey company. Second, it could seek to identify the model that the survey company used to construct the sample that has been provided. Each of these two options is accompanied by a certain perspective concerning how the data should be treated. In the first instance, the bogus sampling scheme used by the survey company is viewed as a nuisance. In the second instance, the viewpoint shifts to one of trying to understand how the sample was constructed, in order to better utilize the information provided.

TABLE 1.1 THE THREE SAMPLING DISTRIBUTIONS FOR THE CONTRIVED ILLUSTRATION

Sample Size	Sample Composition #As	#Bs	#Cs	Number of Occurrences	$\bar{x}$
2	2	0	0	1	2.0
	1	1	0	2	3.0
	1	0	1	2	4.0
	0	2	0	1	4.0
	0	1	1	2	5.0
	0	0	2	1	6.0
weighted 2	5	0	0	1	2.0
	3	2	0	1	2.8
	3	0	2	1	3.6
	2	3	0	1	3.2
	2	0	3	1	4.4
	0	5	0	1	4.0
	0	3	2	1	4.8
	0	2	3	1	5.2
	0	0	5	1	6.0
5	*5	0	0	1	2.0
	4	1	0	5	2.4
	4	0	1	5	2.8
	*3	2	0	10	2.8
	3	1	1	20	3.2
	*3	0	2	10	3.6
	*2	3	0	10	3.2
	2	2	1	30	3.6
	2	1	2	30	4.0
	*2	0	3	10	4.4
	1	4	0	5	3.6
	1	3	1	20	4.0
	1	2	2	30	4.4
	1	1	3	20	4.8
	1	0	4	5	5.2
	*0	5	0	1	4.0
	0	4	1	5	4.4
	*0	3	2	10	4.8
	*0	2	3	10	5.2
	0	1	4	5	5.6
	*0	0	5	1	6.0

*Denotes that the sample composition for n = 5 is also a possible sample composition for the weighted n = 2 case.

Why Autocorrelation is Important to Geographers and Geography

The example outlined is exactly the sort of problem to which spatial autocorrelation refers. A parallel example can be constructed in the following manner. Consider an agricultural experiment station in which a tract of land has been divided into parcels, and randomly selected parcels have had fertilizer applied to them. Because rainfall causes part of this fertilizer to migrate from fertilized parcels to adjacent parcels, the quantity and quality of information from fertilizer experiments is not what it appears to be when superficially studied, just as the aforementioned survey company's sample was not what it appeared to be. The spillover of fertilizer that occurs means that some of the information collected about agricultural yields will be redundant.

An experimenter has three concerns here. First, the suspicion that marked spillover effects exist can be assessed from a probabilistic viewpoint, similar to what the research firm did when it scrutinized the sample it received and found that what the survey company supplied to it was unlikely to occur if pure random sampling had been engaged. In the case of this geographic sample, one wants to know whether or not the geographic distribution of crop yields by parcel is likely to have occurred if no spillovers of fertilizer had taken place. Statistics are available to aid in making this decision (the Moran Coefficient, the Geary Ratio, the Join Counts statistic).

Second, if an experimenter decides that spillovers have occurred, then the functional form of the spillover can be sought. This is equivalent to the research firm discovering that the survey company had weighted the first sampled observation selection by 2 and the second sampled observation selection by 3. In other words, what is the functional form that best describes this spillover process? Here one should recognize that the functional form allows the extent of certain parameter biases to be ascertained. As in the contrived illustration, commonly estimates of the parameters of geographic distributions remain unbiased, whereas spillover effects introduce bias into the standard errors of these parameter estimates. Knowing the functional form will permit a correction of this bias.

Knowing the functional form also reflects upon map pattern, a topic of interest to many geographers. The existence of map pattern means that if information is collected for a given areal unit, then there is an increased chance (better than a fifty-fifty chance) in a forecasting sense that insightful statements can be made about nearby areal units. In terms of the foregoing example, this situation is equivalent to knowing what the first sample selection is, and since it is repeated twice then predicting what the remaining three selections will be. If pure random sampling would be employed, the probability of correctly guessing the remaining sample entries is 1/27. But under the survey company's guidelines, this probability increases to 1/2. In other words, the presence of redundant information in measurements for one areal unit allows somewhat accurate predictions to be made about what seems to be the information content associated with measures made on juxtaposed areal units.

Third, the experimenter can consider spillover effects to be a nuisance, and seek the underlying set of independent elements. This strategy would be equivalent to the research firm attempting to uncover the sampling distribution for a sample size of $n = 2$, which in fact actually is what the survey company utilized. The spillover effects are regarded as being an unimportant model component and hence a nuisance, and as such all redundant information is removed from the sampled observations. Distortions due to spillover effects are discarded. In this way the underlying quantity and quality of information is recognized and assessed.

Therefore, spatial autocorrelation is important to geographers and geography because it reflects upon the quantity and quality of information contained in spatial data, and ultimately then in the soundness of data interpretations.

Tobler (1970) noted that all locations on a geographic surface are related to each other, but closer locations are more strongly related than those more distant. In other words, spatial analysts should expect that the information content of data values for two juxtaposed areal units will contain some redundancies. If test statistics support this contention, then statistical hypothesis testing based upon the raw data will employ an incorrect standard error estimate, often leading to incorrect statistical decisions. With regard to the fertilizer example, spillover effects refer to the migration of fertilizer from a fertilized plot to an unfertilized, adjacent plot. Plots that would be classified as being unfertilized would have their respective yields increased by this fertilizer migration. An increase in yields on plots that did not have fertilizer applied would result in a decrease in the average difference in mean yield for fertilized and unfertilized plots, as well as a reduction in the variation of crop yields over all plots. Thus vastly different levels of fertilizer applications could lead to the false conclusion that fertilizer applications have little effect on crop yields. To avoid this source of incorrect decisions, either the standard errors in question need to be corrected (*i.e.,* the functional form of the spillover effects needs to be identified), or all redundant information must be factored out of the raw data (*i.e.,* the underlying set of independent data values needs to be evaluated).

In terms of the agricultural experiment, the original problem reduces to one of determining whether or not crop yields are strictly a function of fertilizer types and levels. However, if fertilizer spills over from a fertilized parcel to adjacent differently fertilized parcels, such spillovers introduce another source of influence on crop yields that needs to be controlled. Primary goals of spatial autocorrelation analysis include detecting the presence of these sorts of spillovers, and controlling the results of their presence. Thus spatial autocorrelation analysis frequently affords a deeper understanding of geographic landscapes and processes. Moreover, it accounts in part for map pattern. If positive spatial autocorrelation prevails, then visual local geographic differentiation becomes depressed, whereas statistical measures of regional variations tend to become significant. Conversely, if negative spatial autocorrelation prevails, local geographic differentiation becomes conspicuous, while statistical measures of regional variations tend to be insignificant. Properly acknowledging the presence of spatial autocorrelation resolves these inconsistencies.

These are the two classes of reasons stating why autocorrelative structures are important to geographers and geography. The first set of reasons has to do with statistical estimation of parameters for models of spatial data series, whereas the second set has to do with the conceptualization of spatial processes. In the first case, when autocorrelation occurs, difficulties are commonly encountered that normally lie dormant in statistical analysis in general. Foremost is that the sampling distributions for many conventional test statistics, such as t and F, fail to adhere to their theoretical distributions under the null hypothesis because the standard errors are incorrect (affecting Type I error, the probability of rejecting the null hypothesis when it is true, and Type II error, the probability of failing to reject the null hypothesis when it is false). Another complication is that relatively simple estimation methods, such as ordinary least squares, become inappropriate, and must be replaced by far more sophisticated estimation procedures (Ord 1975). An additional problematic feature is that many classical statistical properties of parameter estimates, such as biasness, sufficiency, efficiency, and consistency, are lost. Therefore geographers must both recognize and correctly handle spatial autocorrelation if their statistical analyses are to be useful within an inferential context or to have scientific meaning.

One example of this hindrance to proper statistical decisionmaking was the question of whether or not average levels of agricultural production of eight crops differed across the five regions of Puerto Rico (Griffith 1979). A statistical analysis of the raw data suggested that the answer to this question should be 'yes.' An exploration of the eight geographic distributions of crop yields under study indicated that marked spillover effects were present in each, implying that the raw data contained a considerable amount of redundant information. These spillover effects can be attributed to common local weather patterns, common local terrain features and soil types, common sources of demand for agricultural products (*i.e.,* nodal regions in the Von Thünen sense of commercial agricultural production), patterns of information diffusion (*e.g.,* being aware of what neighboring farmers find profitable to grow), and the like. Removal of these redundancies resulted in an answer to the aforementioned question of 'no.' Thus spillover effects alone caused an incorrect decision to be reached on the basis of the raw data.

The second set of reasons explaining the importance of autocorrelation to geographers and geography is conceptual in nature, particularly in the context of spatial interaction models, to name a few. The argument put forth by Curry (1972), Fotheringham (1981, 1984), Griffith and Jones (1980), and Sheppard (1979) and others is that gravity spatial interaction models overlooking prevailing spatially autocorrelative structure are misspecified. More specifically, the prevailing level of spillover effects that arise, for example, from urban agglomeration forces impacts upon estimated values of distance decay parameters. Regional spatial patterns in distance decay parameters have been documented by Gould (1975). Furthermore, Griffith and Jones (1980) have found that the variation in distance decay parameters over a set of urban places is partially predictable from measured levels of spatial autocorrelation for geographic distributions within urban areas. Consequently, geographers must acknowledge the presence of autocorrelative structures if they are to properly specify their conceptual models. Suppose a spatial interaction model is used to estimate public transit ridership. Misspecification could result in serious errors in predictions of the number of riders between two points. Such errors are serious because they lead to a misallocation of transit vehicles, resulting in certain transit routes having more vehicles than are needed by riders (decreasing benefits and increasing costs of the transit system), while other transit routes will have fewer vehicles than are needed by riders (decreasing benefits of the transit system and, eventually, income).

The Nature of Structure in Scientific Inquiry

The pattern of spillover effects discussed in the previous section was determined by the prevailing arrangement of areal units and, by implication, the underlying geographic distribution of soil types, terrain features, and so on. Interdependence and structure have been focal points of scientific inquiry for decades, if not centuries. *Interdependence* refers to the nature and degree of mutual influences exerted by objects upon one another. *Structure* refers to patterns displayed by these influences, as well as the functional organization of sources of these influences. Until somewhat recently science has tended to analyze interrelationships among variables. Multivariate techniques, for instance, were developed to address questions within a sampling framework that asked whether or not one variable could be predicted from some other variable or variables (regression); whether or not a set of variables measured different features (principal components or factor analysis); whether or not observations could be classified into distinct groups (univariate or multivariate analysis of variance, or discriminant function analysis); and whether or not sets of variables were measures of similar characteristics (canonical correlation). The class of geographical problems in this monograph shares the same two points of focus, namely interdependence and structure, but in a rather specialized way. Interrelationships will be explored that involve areal observational units. These areal units may refer to either the points on a punctiform surface, or the partitionings of a planar surface that has been divided into mutually exclusive and collectively exhaustive polygons. Hence observational interrelationships arising from the manner in which areal units are arranged on a planar surface will be explored in terms of interdependence and structure. The nature and degree of these interrelationships have become known as spatial autocorrelation.

The Ways Things are Structured

Berry (1964) formalized the notion of a geographic matrix, where the rows generally denoted observations, and the columns generally had attribute labels. He then noted that designating areal units as the observations added a geographic dimension to the data in question. Studies that focus on a single column attribute are concerned with spatial variation, while studies that focus on several columns are concerned with areal associations. Studies that focus on a single row areal unit are concerned with place inventories, and studies that focus on a number of rows tend to be concerned with regionalization (Figure 1.1).

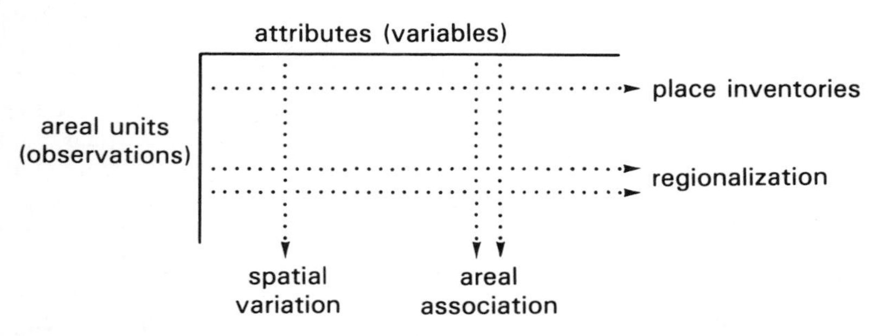

FIGURE 1.1 Berry's Geographic Matrix (after Berry 1964)

The notion of structure refers to three different features of this geographic matrix. One type of pattern commonly studied in scientific inquiry is a conspicuous pattern of covariation among the columns of the matrix. In other words, the structure exhibited by a set of variables is the focal point of interest. Another pattern commonly studied is the similarity of observations across a set of attributes. Here the structure exhibited by a set of observations is the focal point of interest. Both of these traditional perspectives on structure are accommodated and explicated to some degree in Berry's presentation, and depend on areal units on a map. Given this matrix, can the arrangement of areal units be specified? The map from which a geographic data matrix has been constructed *cannot* be retrieved from the matrix.

The importance of the property of spatial arrangement is the theme of this monograph. Because such phenomena as the migration of fertilizer from fertilized parcels to adjacent parcels occur, recognition of areal unit juxtaposition furnishes a better understanding of many spatial processes. Consequently, the phrase "adding a spatial dimension to a problem" takes on a new meaning. Moreover, the spatial dimension will suggest that interdependence and structure arise from absolute locations for areal units (*e.g.,* latitude and longitude coordinates), relative locations for areal units (*e.g.,* the nearest neighbor arrangement), and the intensity of interrelationships among phenomena residing in these areal units. Coupled with Berry's geographic matrix conception, it provides the necessary basis needed to analyze spatial distributions as spatial data series.

Generally speaking, because geography is a synthesizing discipline, many of its developments have tended to parallel those of its cognate disciplines. Accordingly, subsequent discussion frequently will refer to parent developments in classical statistics and time series analysis.

Autocorrelation in Time Series Versus Spatial Series

Much of the early work in spatial autocorrelation analysis received its impetus from and paralleled developments that occurred historically in time series analysis. Certain statistics, such as the Runs statistic, were found to generalize quite easily to two-dimensional situations. However, much recent work in this research area eloquently demonstrates that difficulties are commonly encountered in the analysis of spatial series that normally lie dormant in the analysis of time series. One of the most conspicuous examples of this sort of difference again is found in the applicability of the ordinary least squares procedure to time series data, but not to spatial series data. It will be shown that since spatial data embrace bidirectional dependencies on a two-dimensional surface that are in some sense reciprocal in nature, considerable complications arise in statistical analysis of such data that often prevent traditional procedures such as ordinary least squares from being applicable. In contrast, time series data embrace unidirectional dependencies along one-dimension that obviously are not reciprocal in any way. Hence the complications introduced into time series analysis tend to be substantially less complicated, and usually allow traditional statistical procedures to be applied.

2

Defining Spatial Autocorrelation

In the classical sense correlation refers to the relationship prevailing between a pair of variables. Coefficients used in conventional statistics to measure this relationship provide information about its nature and degree. The sign of a correlation coefficient indicates the nature of the relationship; a plus identifies a positive or direct relation, while a minus identifies a negative or inverse relation. The actual numerical value identifies the strength of a given relation. Let ρ denote the correlation coefficient. Then, $|\rho| \rightarrow 1$ implies an increasingly strong relation, whereas $|\rho| \rightarrow 0$ implies an increasingly weak relation.

The selection of a correlation coefficient to measure some relationship depends upon the measurement scales of the two variables involved. If both variables are measured at the nominal scale, then a phi coefficient is used. If either is measured at the nominal scale while the other is measured at the interval/ratio scale, then a point biserial coefficient is used. If both variables are measured at the ordinal scale, then Spearman's rank correlation coefficient is used. Finally, if both are measured at the interval/ratio scale, Pearson's product moment correlation coefficient is used.

What Is Meant by Autocorrelation

Now that a definition of correlation has been established, the "auto-" part of autocorrelation needs to be clarified. This prefix means self; hence, autocorrelation may be loosely defined as self–correlation. As such it involves a single variable, and refers to the correlation between pairs of observations made on this single variable. This type of correlation arises because realizations of the variable in question are ordered in some way. Thus, autocorrelation may be defined as the relationship among values of some variable that is attributable to some underlying ordering of these values. This notion is at odds with classical statistical analysis, which assumes that such an ordering does not exist, and hence observations always are pairwise independent (Gould 1970).

The ordering mentioned above can be described as a configuration for the n observed values of some variable X, and may be depicted by an n-by-n table or matrix of values c_{ij}, say **C**, whose row and column labels are the same sequence of observations. Zero entries in this matrix would indicate that the corresponding row and column observations are not side-by-side, or in juxtaposition with one another, in the associated ordering. Unity entries in this matrix would indicate that the corresponding row and column areal units are juxtaposed. Consider an ordering that refers to a sequence through time, where realizations x_t and x_{t+1} may be related (Figure 2.1). If n observations are ordered in such a way that the time subscript is ascending, then all entries in the configuration matrix would be zero, except for the upper off-diagonal cells, which would be unity. Accordingly:

$$\mathbf{C} = \begin{matrix} 0 & 1 & 0 & 0 & \cdots & 0 & 0 \\ 0 & 0 & 1 & 0 & \cdots & 0 & 0 \\ 0 & 0 & 0 & 1 & \cdots & 0 & 0 \\ \cdot & \cdot & \cdot & \cdot & \cdots & \cdot & \cdot \\ \cdot & \cdot & \cdot & \cdot & \cdots & \cdot & \cdot \\ \cdot & \cdot & \cdot & \cdot & \cdots & \cdot & \cdot \\ 0 & 0 & 0 & 0 & \cdots & 0 & 1 \\ 0 & 0 & 0 & 0 & \cdots & 0 & 0 \end{matrix}$$

Only the upper off-diagonal elements are unity because temporal influences are unidirectional. One of the closest geographic analogies to this particular situation occurs in river networks, where water flows only in one direction. As a result, time series methodology has been used for the statistical analysis of river networks.

If a transect spatial sample through some region is selected, though, then all entries of matrix **C** would be zero except for both the upper and lower off-diagonals, which would be unity. Accordingly:

$$\mathbf{C} = \begin{matrix} 0 & 1 & 0 & 0 & \cdots & 0 & 0 \\ 1 & 0 & 1 & 0 & \cdots & 0 & 0 \\ 0 & 1 & 0 & 1 & \cdots & 0 & 0 \\ \cdot & \cdot & \cdot & \cdot & \cdots & \cdot & \cdot \\ \cdot & \cdot & \cdot & \cdot & \cdots & \cdot & \cdot \\ \cdot & \cdot & \cdot & \cdot & \cdots & \cdot & \cdot \\ 0 & 0 & 0 & 0 & \cdots & 0 & 1 \\ 0 & 0 & 0 & 0 & \cdots & 1 & 0 \end{matrix}$$

Here the same visual configuration of observations exists, as a comparison of Figures 2.1a and 2.1b illustrates. But geographic influences are bidirectional. In addition, the set of areal units can occur in two dimensions, not just one, as Figure 2.1c demonstrates. This extension to two dimensions will result in matrix **C** having unity entries in many more than just the off-diagonal positions.

Given these notions of autocorrelation, ordering, and configuration, attention now will be turned to defining spatial autocorrelation.

The Phenomenon of Spatial Autocorrelation

The term autocorrelation found in the phrase "spatial autocorrelation" is defined no differently than before. The term spatial refers to the way in which values of some variable X are ordered. Therefore, the ordering in question describes how areal units are arranged on a planar map. Moreover, here unity entries in the configuration matrix **C** denote which areal units are in juxtaposition with one another on a two-dimensional surface.

Clearly, if spatial data are collected at points, then these data points cannot

(a) A Time Series Configuration

X_j	X_{j+1}	X_{j+2}	X_{j+3}	X_{j+4}	X_{j+5}

(b) A Linear Spatial Series Configuration

$X_{i,j}$	$X_{i,j+1}$	$X_{i,j+2}$
$X_{i-1,j}$	$X_{i-1,j+1}$	$X_{i-1,j+2}$
$X_{i-2,j}$	$X_{i-2,j+1}$	$X_{i-2,j+2}$

(c) A Two-dimensional Spatial Series Configuration

FIGURE 2.1 Selected Orderings of Observations

be side-by-side in the physical sense of the phrase. Rather, unity entries in configuration matrix **C** will indicate which dyads of points are relatively close. The proximity of pairs of points may be translated into an ordering over a planar surface in several different ways. The two most preferable of these translations are described by Ripley (1981) and Matula and Sokal (1980). In the first of these two cases, the surface is partitioned into a set of Thiessen polygons, generated from the initial point pattern. Then, the configuration matrix **C** is constructed so that unity entries in it indicate that the corresponding row and column points have common sides of Thiessen polygons. This approach is illustrated by Griffith (1979), who used it in his analysis of grain elevator data for the Canadian Prairies.

A Thiessen polygon is constructed by connecting all nearest neighbor points on a surface with straight lines and then constructing perpendicular bisectors of these connections (Figure 2.2). Points where these bisectors meet constitute the vertices of the polygons. The attractive feature of Thiessen polygons is that all points on the geographic surface contained within a polygon's boundaries are closest to the given polygon's spatial data point than to any other spatial data point on the surface. Also displayed in Figure 2.2 is the resulting connectivity matrix for the sample set of spatial data points. A computer algorithm for constructing these sorts of polygons has been devised by Brassel and Reif (1979).

Therefore, spatial autocorrelation may be defined as the relationship among values of some variable that is attributable to the manner in which the corresponding areal units are ordered or arranged on a planar surface. The associated configuration matrix **C** depicting this ordering will have entries of zero except where polygon type areal units are in juxtaposition with one another, or except where links exist in some graph theoretic representation of

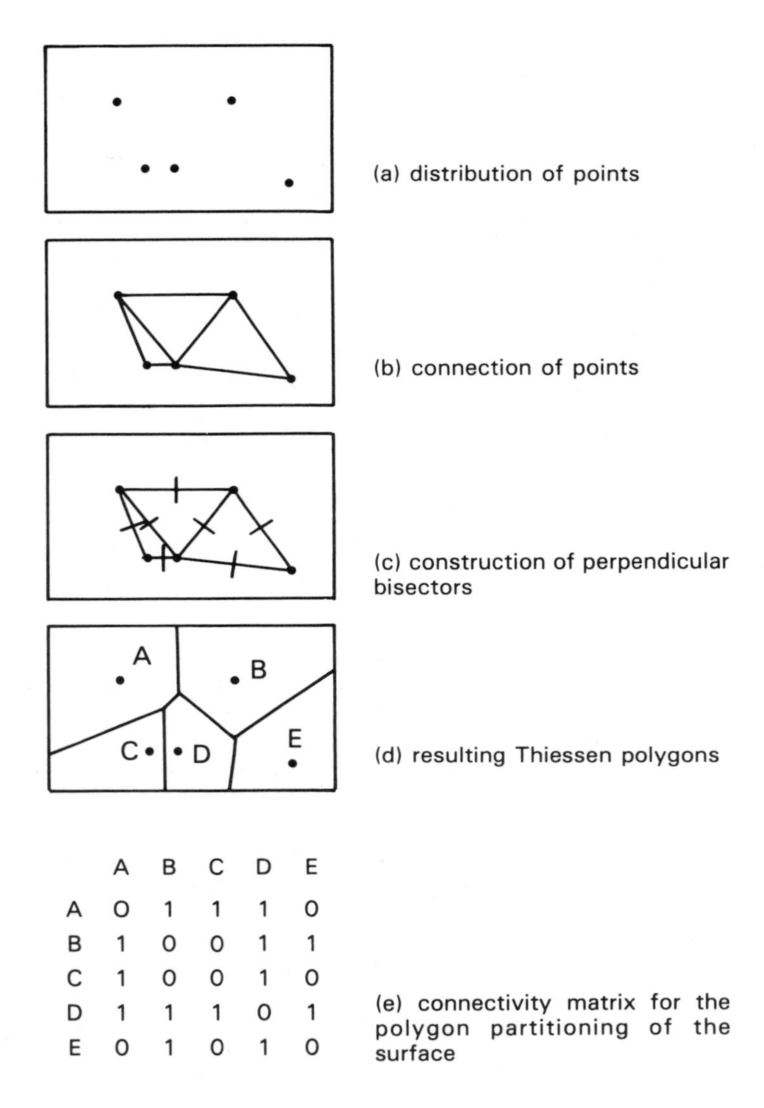

	A	B	C	D	E
A	0	1	1	1	0
B	1	0	0	1	1
C	1	0	0	1	0
D	1	1	1	0	1
E	0	1	0	1	0

FIGURE 2.2 Constructing a Thiessen Polygon Partitioning of a Punctiform Surface

point patterns. In either of these latter situations the matrix **C** entry will be unity. An important property of this matrix is that each row/column entry is identical to its corresponding column/row entry (symmetry). Analogous to the case of classical correlation, different spatial autocorrelation coefficients have been derived where variable X is measured on a nominal scale (Dacey 1968), on an ordinal scale (Sen and Soot 1977), or on an interval/ratio scale (Moran 1948; Geary 1954).

Sources of Spatial Autocorrelation

Spatial autocorrelation has two primary connotations as a statistical property of spatial data sets (Haining 1980; Haining, Griffith and Bennett 1983), based on the arrangement of areal units or on the underlying probability distribution of spatial values. Most often spatial autocorrelation has been viewed as a property of the arrangement of n areal units upon a planar surface. Suppose there exists a set of n values of some variable X, say $\{x_i;\ i=1,2,\ldots,n\}$. One observes a spatial distribution in which each of these n values is affiliated with one of the n areal units. In tabular form this correspondence between values x_i and areal units may be represented by entering the areal unit names as the row labels of a table, and then entering the respective x_i values in those cells produced by intersecting these rows with a column of this table that is identified by the name of the variable in question.

One question that may be asked at this point has to do with the structure of an observed set of values, and may be phrased as follows:

> if the set of variable values $\{x_i\}$ had been allocated in a random fashion to the set of n areal units appearing on the two-dimensional map, would this observed spatial distribution be a likely outcome of such an allocation procedure?

In order to answer this particular question, four features of this problem need to be recognized. First, it is necessary to define a statistic that can discriminate between different arrangements of the set $\{x_i\}$ on a two-dimensional surface. These kinds of measures will be discussed later.

Second, the different arrangements that are possible need to be identified, so that the appropriate sampling distribution can be constructed. A first step in such an identification is the counting of these possible arrangements. There exist n values and n areal units. Once a value has been allocated to an areal unit, it is not available for allocation to any of the remaining areal units. To begin with then, there are n choices for the first areal unit. After selection of an x_i value, there remain (n-1) choices for the second areal unit. After selection of one of the remaining available x_i values for this areal unit, there remain (n-2) choices for the third areal unit. This selection procedure continues until all values of x_i have been allocated, which is equivalent to each areal unit having had an x_i value allocated to it. Since there are n possibilities for the first choice, and (n-1) possibilities for the second choice, then there are n(n-1) possibilities for the first and second choices together. Extending this counting principle for all n areal units results in n! possible spatial distributions; hence, one is looking at all possible permutations of the set $\{x_i\}$ over the map, a situation that is analogous to sampling without replacement, where order is important.

The aforementioned counting principle assumes that the set $\{x_i\}$ consists of n distinct values. However, suppose $x_1 = x_2$ and in one allocation x_1 is assigned to areal unit A while x_2 is assigned to areal unit B. If these two assignments are switched with one another in a new allocation, the spatial distribution would remain unchanged. Suppose there are $k < n$ sets of distinct values; let n_j denote the number of times each distinct value appears in the set. Combinatorial theory states that these redundant, or repetitive, map allocations can be eliminated from the enumeration count by dividing n! by $\prod_{j=1}^{j=n} (n_j!)$. Consequently, the total number of allocations becomes $n!/\prod_{j=1}^{j=n} (n_j!)$. One should note that when all the values are distinct, and hence $k = n$, then $n_j = 1$ for all j, and $\prod_{j=1}^{j=n} (1!) = 1$, resulting in the reduction of this adjusted expression to n!.

A third feature builds upon this permutation viewpoint, which is equivalent to an experimental approach using randomization. This feature has to do with whether or not the allocation that is observed is representative of one resulting from a random allocation of the set $\{x_i\}$ to the n areal units. The statistic needed to measure spatial autocorrelation in this context, then, should have a sample mean that corresponds to zero autocorrelation. Furthermore, critical values for it need to be established in order to determine which spatial distributions of the x_i values are not representative of a random allocation. This classification of statistic values into the two groups of representative and non-representative (of zero spatial autocorrelation) is consistent with the establishment of 10%, 5% and 1% levels of significance in classical statistics.

A fourth salient feature refers to the nature of the map pattern that could be observed. On the one hand, if similar x_i values have a high propensity to cluster on the map, exceeding that due merely to chance, then the statistic in question should be indicative of positive spatial autocorrelation. On the other hand, if dissimilar x_i values have a high propensity to cluster, then negative spatial autocorrelation should be detected. Similarity of values in this sense is relative. It refers to ordering x_i in ascending order. If those x_i values that appear at the beginning of this ordering tend to occupy juxtaposed areal units, those that appear in the middle of this ordering tend to occupy juxtaposed areal units, and those that appear at the end of this ordering tend to occupy juxtaposed areal units, then as the value of x_i increases the values of the surrounding x_i tend to increase. Hence, *positive spatial autocorrelation* exists in the corresponding map pattern. In contradistinction, if alternating areal units are allocated, in descending order, those x_i values that appear in the first half of the aforementioned ordering, while the remaining areal units (which is every other one) are allocated, in ascending order, those x_i values that appear in the last half of this ordering, then as the value of x_i increases, the values of the surrounding x_i tend to decrease. Hence, *negative spatial autocorrelation* exists in the corresponding map pattern.

The second connotation for spatial autocorrelation refers to the probability distribution that underlies a spatial distribution. As such, a random variable X is assumed to exist for each areal unit. The n variables are assumed to have the same mean, μ, the same variance, σ^2, and the same frequency distribution, say a normal distribution. Next, the observed map pattern is considered to be a random selection from a multivariate distribution. Consequently, if zero spatial autocorrelation is present among the random variables X_i, then a correlation matrix **R** for these variables, based upon data for the population rather than a sample, would be equal to the identity matrix, where $\rho_{ii} = 1$ and $\rho_{ij} = 0$ $(i \neq j)$. Otherwise, $\rho_{ij} \neq 0$ $(i \neq j)$ when areal units i and j are juxtaposed.

Once again, if $\rho_{ij} > 0$ then similar x_i values would tend to cluster on the map. If $\rho_{ij} < 0$ then dissimilar x_i values would tend to cluster on the map. This perspective is overshadowed by the inability to observe more than one surface, or more than one realization of variable X, unless a simulation experiment is being performed or a space-time series is being studied. Accordingly, in practice a sample of size one usually must be used to evaluate the nature and degree of spatial autocorrelation. Because of this constraint, the assumption of identically distributed values of X_i is a fundamental requirement for obtaining analytical results. If a large number of two-dimensional surfaces can be observed, which are representative samples of a random relation between values of X_i, then the average level of spatial autocorrelation measured for these surfaces should be approximately zero. As the number of observed surfaces increases to infinity, this average value should converge upon zero. Hopefully, when only one spatial distribution is observed, it is representative of the total number of possible spatial distributions.

This time the question that may be asked has to do with the joint distribution of the X_i variates, and may be phrased as follows:

> if the n realizations of the random variables X_i are mutually independent, is the single sample of joint realizations constituting the observed map pattern a representative one?

To answer this question, several features of the problem need to be recognized. First, it is again necessary to define a statistic that can discriminate between different natures and degrees of spatial autocorrelation. In addition, the underlying spatial distribution needs to be tested for. If a finite number of two-dimensional realizations is available, such as in a simulation experiment, each set of areal unit X values can be tested. If only one map pattern is available, though, the assumption of identically distributed X_i values needs to be invoked. By doing so, the set of observed x_i values can be pooled into a single frequency distribution, and then this frequency distribution can be tested. Evidence exists to suggest that this latter test may be erroneous if the parent correlation matrix is not the identity matrix (Griffith 1980).

Simulation Experiments Employing the Randomization Connotation

The first of the two connotations for spatial autocorrelation is the easier one to use to illustrate the phenomenon itself. A BASIC computer program that was developed for the PC-compatible microcomputer appears in Appendix A, a tool for conducting laboratory simulation experiments regarding spatial autocorrelation. A set of numbers that distributed over a given map must be rearranged to achieve some pre-specified nature and degree of spatial autocorrelation. As the numbers are sequentially rearranged, a spatial autocorrelation coefficient is calculated and displayed so that the student can monitor his/her progress toward the desired coefficient.

The simulation experiment is for a 4-by-4 regular lattice configuration of areal units (Figure 2.3). A set of 16 integral numbers, between 10 and 99, is selected at random such that the probability of selecting each number is 1/90. The minutes of the computer clock are used to select these numbers, resulting in 60 different possible geographic distributions. Then the 16 sampled numbers are allocated to the 4-by-4 lattice in a random fashion. The simulation experiment involves the following four steps:

Step 1: note the initial geographic distribution, which is randomly constructed by the computer, together with its affiliated spatial autocorrelation coefficient;

Step 2: systematically rearrange the selected set of numbers until a high degree of positive spatial autocorrelation is present (*i.e.*, the coefficient is close to unity), and record the resulting geographic distribution;

Step 3: systematically rearrange the new geographic distribution of numbers resulting in Step 2 until virtually no spatial autocorrelation is observed (*i.e.*, the coefficient is close to zero), and record the resulting geographic distribution; and,

Step 4: systematically rearrange the new geographic distribution of numbers resulting in Step 3 until a high degree of negative spatial autocorrelation is present (*i.e.*, the coefficient is close to minus one), and record the resulting geographic distribution.

Finally the student should be requested to compare those limiting cases map patterns constructed in Steps 2, 3, and 4. The student also should calculate the number of possible map patterns (a maximum of $16! \approx 2.09 \times 10^{13}$).

Figure 2.3 displays the results of a sample run of this preceding simulation experiment. The computer selected as the set of 16 random numbers {13,16,19,23,26,27,38,38,41,50,55,58,75,78,83,84}. For illustrative purposes these numbers might be thought of as the number of dentists per 10,000 inhabitants. The initial random allocation of these values to the corresponding map is given in Figure 2.3a. The observed level of spatial autocorrelation here is measured as -0.0440, which is only slightly different from a value of zero, denoting no geographic dependency. In terms of the dentists example, this pattern would indicate that dentists tend to locate at random. Reorganizing these sixteen numbers such that large values are in juxtaposition with one another, and small values are in juxtaposition with one another, suggests that the set of values {75,87,83,84} should cluster on the map, as should {13,16,19,23}. A permutation of the original map pattern to that given in Figure 2.3b achieved this goal, and yielded an autocorrelation measure of 0.7526, which is reasonably close to unity (a perfect direct relationship). With regard to the dentists example, this pattern would indicate that significant regional differences exist in the supply of dentists. Similarly, reorganizing these sixteen numbers such that large values are in juxtaposition with small values suggests that, for instance, 84 needs to be surrounded by the set of values {13,16,19,23} on the map. A permutation of the original map pattern to that given in Figure 2.3d achieved this goal, and yielded an autocorrelation measure of -0.8952, which is reasonably close to minus one (a perfect indirect relationship). Within the context of the dentists example, this pattern would indicate that high numbers of dentists in one area draw patients

(a) Computer Generated Map

16 A	78 E	83 I	13 M
38 B	50 F	58 J	23 N
55 C	41 G	38 K	75 O
19 D	84 H	26 L	27 P

spatial autocorrelation coefficient −0.0440

(b) Permutation Heuristically Maximizing Positive Spatial Autocorrelation

84 A	83 E	58 I	38 M
78 B	75 F	50 J	27 N
55 C	41 G	26 K	19 O
38 D	23 H	16 L	13 P

spatial autocorrelation coefficient 0.7526

(c) Permutation Heuristically Minimizing Spatial Autocorrelation

19 A	83 E	84 I	13 M
38 B	55 F	58 J	26 N
50 C	41 G	38 K	75 O
16 D	78 H	23 L	27 P

spatial autocorrelation coefficient 0.0012

(d) Permutaion Heuristically Maximizing Negative Spatial Autocorrelation

50 A	23 E	58 I	38 M
19 B	84 F	16 J	55 N
78 C	13 G	83 K	27 O
38 D	75 H	26 L	41 P

spatial autocorrelation coefficient −0.8952

FIGURE 2.3 Sample Output From the Spatial Autocorrelation Program (Appendix A)

from adjoining areas, which therefore have fewer dentists.

This type of simulation experiment helps furnish the student with a feel for what the concept of spatial autocorrelation refers to. In the example presented here there are 16!/2! possible map patterns to choose from (since 38 repeats itself), and the student is seeking various extreme patterns. Further skill development in comprehending the meaning of spatial autocorrelation is attained when the student attempts to produce a completely random pattern (Figure 2.3c). In this particular case sometimes large values are juxtaposed as well as sometimes large values are adjacent to small values. Trying to obtain a coefficient of zero requires making trade-offs between these two tendencies, and this further enhances the understanding of what spatial autocorrelation means. In terms of the dentists example, this pattern would be interpreted the same way as was the one in Figure 2.3a.

3

Spatial Autocorrelation and Geographic Samples

Many researchers are interested in drawing general conclusions from their findings, and hence wish to work with randomly sampled data rather than data constituting a restricted population or data sampled in such a way that only sample descriptions are possible. This desire on the part of geographers has led to Upton and Fingleton's (1985:325) observation that "... great imagination has gone into turning what appears to be a population into a sample." To this end, in spatial autocorrelation two popular perspectives have been employed. One has to do with treating the order of arrangement of data over a geographic surface as important, and then assuming that the observed map is a single sample from the set of all possible data permutations for the given map. The second perspective essentially has to do with stratified random sampling, where the areal units of a map represent a stratification, and data for each areal unit are treated as a random sample from a population in that areal unit. Geographic populations may be defined in a number of other ways (see Griffith 1984; Summerfield 1983). Only the two types mentioned here will be discussed in this monograph, namely and respectively those associated with randomization and those associated with joint normality.

The Assumption of Randomization

In one of the preceding sections a description was provided for the idea of enumerating all possible permutations of a set of n values $\{x_i\}$ over a set of n areal units. Since a population is the total set of items for which a measurement is to be obtained, the set of $n!/\prod_{j=1}^{j=k}(n_j!)$ spatial distributions forms the geographic population being studied. Clearly this population is finite, even though for large numbers of areal units the size of the population will be considerable. A statistical population is the set of measures corresponding to each entry in the population. Thus, the statistical population here will consist of the set of $n!/\prod_{j=1}^{j=k}(n_j!)$ spatial autocorrelation statistics, one for each map pattern in the parent population. The associated spatial sampling problem, then, involves randomly selecting one or more of the possible map patterns. In practice, a single spatial distribution is preselected, namely the observed map pattern.

The other type of geographic population alluded to in the previous section is analogous to classical stratified random sampling. A sample space is constructed for each of the areal units. The total number of possible realizations of variable X_i may be denoted by N_i. If all the values of N_i are finite, and a random sample of size m_i is selected in each areal unit, then there are $(N_i)^{m_i}$ possible samples of this size. Since the samples for each areal unit are being selected in an independent fashion, the total number of spatial distributions is $\prod_{i=1}^{i=n} (N_i)^{m_i}$. The statistical population in this case would consist of the set of spatial autocorrelation statistics that are derived from these $\prod_{i=1}^{i=n} (N_i)^{m_i}$ spatial distributions. The associated spatial sampling problem, then, involves randomly selecting a sample of size m_i in each of the n areal units. If an underlying probability distribution is assumed, then the frequency distribution for each of these n sets of values $\{x_{ik}; k=1,2, \ldots ,m_i\}$ can be tested against the assumed distribution.

Rather than a stratified sample situation, the map pattern may be viewed as the outcome of some random process operating within each areal unit. Now the population is infinite; only one realization of these joint random processes is available for evaluation; spatial sampling consists of each joint realization; and most analyses are based upon the assumed underlying multivariate probability density function. A test of the marginal distributions for this density function can be made by pooling the observed set of x_i values. Usually the multivariate normal density function is assumed.

The Assumption of Normality

Rather than enumerating all possible permutations of a set of n values $\{x_i\}$ over a set of n areal units, as was mentioned earlier, one could consider a map pattern as being the joint realization of n random variables X_i. In order to derive analytical results for this situation, the theoretical frequency distribution, or the probability density function, of these X_i values must be known. Because of the richness of normal curve theory, often these X_i values are assumed to be multivariate normal. This particular assumption allows many illuminating results to be obtained that parallel those that are summarized in multivariate statistical theory.

Frequently the assumption of multivariate normality has a much sounder basis, though. Recalling that when random samples of size m_i are selected from areal unit populations of size N_i, then the total number of possible spatial distribution is $\prod_{i=1}^{i=n} (N_i)^{m_i}$. If some attribute is measured on these $\sum_{i=1}^{i=n} m_i$ observations, and say each areal unit mean $\bar{x}_i$ is calculated for this attribute, then the central limit theorem states that the distribution of each $\bar{x}_i$ is asymptotically normal. Consequently, the joint distribution of a set of $\bar{x}_i$ values would be multivariate normal, since each of the marginal distributions is univariate normal with the same mean μ and variance σ^2.

This property holds for all situations involving sampling. Even if the underlying frequency distribution of X in any given areal unit is not normally distributed, usually by the time $m = 3$ the sampling distribution of $\bar{x}_i$ does not deviate noticeably from a normal distribution. Therefore, the assumption of normality may be utilized anytime measures of central tendencies are calculated for randomly selected observations. In these sorts of situations inferences are being drawn about the spatial distribution of the population parameters μ_i.

Testing for Normality

One bothersome problem encountered in spatial statistical analyses is that sample sizes are quite small. When dealing with frequency distribution comparisons, this problem often rules out use of the χ^2 statistic, and in some cases may even rule out use of the Kolmogorov-Smirnov statistic. Fortunately, Shapiro and Wilk (1965) developed a statistic specifically for the small sample case of comparing an observed frequency distribution with the theoretical normal distribution. Their statistic is applicable to sample sizes ranging from 3 to 50. Clearly, by the time 50 areal units are involved, problems affiliated with the use of χ^2 or Kolmogorov-Smirnov will have dissipated.

The Shapiro-Wilk statistic is relatively simple to calculate. First, rank the x_is in descending order. Second, calculate the paired differences $|x_i - x_{n-i+1}|$. If n is even, then there are n/2 pairs. If n is odd, then there are (n+1)/2 pairs, where the last pair involves subtracting the mid-point x_i value from itself. Third, a set of coefficients a_i are attached to these differences, and the following sum is calculated:

$$b = \sum_{i=1}^{i=k} a_i |x_i - x_{n-i+1}|$$

where k = n/2 or k = (n+1)/2, depending upon whether n is even or odd. The values for a_i are presented here in Table 3.1. Fourth, the test statistic is calculated:

$$W = b^2 / \sum_{i=1}^{i=n} (x_i - \bar{x})^2$$

As $W \rightarrow 1$, the observed frequency distribution increasingly conforms to a normal distribution; as $W \rightarrow 0$, the observed frequency distribution increasingly deviates from a normal distribution. Table 3.2 presents critical values of W for 10%, 5% and 1% levels of significance.

The Assumption of Identically Distributed Variates

If an underlying probability distribution is assumed, it must be the same for the variate X in all areal units. The assumption of identically distributed X values is employed so that a tractable joint probability density function can be constructed. Further, since only one observed surface is available, any other assumption would be difficult to evaluate.

The researcher wants to be able to say that spatial autocorrelation is detected because either it exists in the population under study, or a non-representative sample has been selected in which detected autocorrelation is attributable to sampling error. If X were to have a different frequency distribution within each areal unit, then these differences could be what is detected, rather than the presence or absence of spatial autocorrelation. Even when the general nature of these frequency distributions is the same, say a normal distri-

TABLE 3.1 COEFFICIENTS a_i FOR THE SHAPIRO-WILK TEST FOR NORMALITY

i \ n	2	3	4	5	6	7	8	9	10
1	0.7071	0.7071	0.6872	0.6646	0.6431	0.6233	0.6052	0.5888	0.5739
2	—	.0000	.1677	.2413	.2806	.3031	.3164	.3244	.3291
3	—	—	—	.0000	.0875	.1401	.1743	.1976	.2141
4	—	—	—	—	—	.0000	.0561	.0947	.1224
5	—	—	—	—	—	—	—	.0000	.0399

i \ n	11	12	13	14	15	16	17	18	19	20
1	0.5601	0.5475	0.5359	0.5251	0.5150	0.5056	0.4968	0.4886	0.4808	0.4734
2	.3315	.3325	.3325	.3318	.3306	.3290	.3273	.3253	.3232	.3211
3	.2260	.2347	.2412	.2460	.2495	.2521	.2540	.2553	.2561	.2565
4	.1429	.1586	.1707	.1802	.1878	.1939	.1988	.2027	.2059	.2085
5	.0695	.0922	.1099	.1240	.1353	.1447	.1524	.1587	.1641	.1686
6	0.0000	0.0303	0.0539	0.0727	0.0880	0.1005	0.1109	0.1197	0.1271	0.1334
7	—	—	.0000	.0240	.0433	.0593	.0725	.0837	.0932	.1013
8	—	—	—	—	.0000	.0196	.0359	.0496	.0612	.0711
9	—	—	—	—	—	—	.0000	.0163	.0303	.0422
10	—	—	—	—	—	—	—	—	.0000	.0140

i \ n	21	22	23	24	25	26	27	28	29	30
1	0.4643	0.4590	0.4542	0.4493	0.4450	0.4407	0.4366	0.4328	0.4291	0.4254
2	.3185	.3156	.3126	.3098	.3069	.3043	.3018	.2992	.2968	.2944
3	.2578	.2571	.2563	.2554	.2543	.2533	.2522	.2510	.2499	.2487
4	.2119	.2131	.2139	.2145	.2148	.2151	.2152	.2151	.2150	.2148
5	.1736	.1764	.1787	.1807	.1822	.1836	.1848	.1857	.1864	.1870
6	0.1399	0.1443	0.1480	0.1512	0.1539	0.1563	0.1584	0.1601	0.1616	0.1630
7	.1092	.1150	.1201	.1245	.1283	.1316	.1346	.1372	.1395	.1415
8	.0804	.0878	.0941	.0997	.1046	.1089	.1128	.1162	.1192	.1219
9	.0530	.0618	.0696	.0764	.0823	.0876	.0923	.0965	.1002	.1036
10	.0263	.0368	.0459	.0539	.0610	.0672	.0728	.0778	.0822	.0862

11	0.0000	0.0122	0.0228	0.0321	0.0403	0.0476	0.0540	0.0598	0.0650	0.0697
12	—	—	.0000	.0107	.0200	.0284	.0358	.0424	.0483	.0537
13	—	—	—	—	.0000	.0094	.0178	.0253	.0320	.0381
14	—	—	—	—	—	—	.000	.0084	.0159	.0227
15	—	—	—	—	—	—	—	—	.0000	.0076

i \ n	31	32	33	34	35	36	37	38	39	40
1	0.4220	0.4188	0.4156	0.4127	0.4096	0.4068	0.4040	0.4015	0.3989	0.3964
2	.2921	.2898	.2876	.2854	.2834	.2813	.2794	.2774	.2755	.2737
3	.2475	.2463	.2451	.2439	.2427	.2415	.2403	.2391	.2380	.2368
4	.2145	.2141	.2137	.2132	.2127	.2121	.2116	.2110	.2104	.2098
5	.1874	.1878	.1880	.1882	.1883	.1883	.1883	.1881	.1880	.1878
6	0.1641	0.1651	0.1660	0.1667	0.1673	0.1678	0.1683	0.1686	0.1689	0.1691
7	.1433	.1449	.1463	.1475	.1487	.1496	.1505	.1513	.1520	.1526
8	.1243	.1265	.1284	.1301	.1317	.1331	.1344	.1356	.1366	.1376
9	.1066	.1093	.1118	.1140	.1160	.1179	.1196	.1211	.1225	.1237
10	.0899	.0931	.0961	.0988	.1013	.1036	.1056	.1075	.1092	.1108
11	0.0739	0.0777	0.0812	0.0844	0.0873	0.0900	0.0924	0.0947	0.0967	0.0986
12	.0585	.0629	.0669	.0706	.0739	.0770	.0798	.0824	.0848	.0870
13	.0435	.0485	.0530	.0572	.0610	.0645	.0677	.0706	.0733	.0759
14	.0289	.0344	.0395	.0441	.0484	.0523	.0559	.0592	.0622	.0651
15	.0144	.0206	.0262	.0314	.0361	.0404	.0444	.0481	.0515	.0546
16	0.0000	0.0068	0.0131	0.0187	0.0239	0.0287	0.0331	0.0372	0.0409	0.0444
17	—	—	.0000	.0062	.0119	.0172	.0220	.0264	.0305	.0343
18	—	—	—	—	.0000	.0057	.0110	.0158	.0203	.0244
19	—	—	—	—	—	—	.0000	.0053	.0101	.0146
20	—	—	—	—	—	—	—	—	.0000	.0049

TABLE 3.1 (Continued)

i \ n	41	42	43	44	45	46	47	48	49	50
1	0.3940	0.3917	0.3894	0.3872	0.3850	0.3830	0.3808	0.3789	0.3770	0.3751
2	.2719	.2701	.2684	.2667	.2651	.2635	.2620	.2604	.2589	.2574
3	.2357	.2345	.2334	.2323	.2313	.2302	.2291	.2281	.2271	.2260
4	.2091	.2085	.2078	.2072	.2065	.2058	.2052	.2045	.2038	.2032
5	.1876	.1874	.1871	.1868	.1865	.1862	.1859	.1855	.1851	.1847
6	0.1693	0.1694	0.1695	0.1695	0.1695	0.1695	0.1695	0.1693	0.1692	0.1691
7	.1531	.1535	.1539	.1542	.1545	.1548	.1550	.1551	.1553	.1554
8	.1384	.1392	.1398	.1405	.1410	.1415	.1420	.1423	.1427	.1430
9	.1249	.1259	.1269	.1278	.1286	.1293	.1300	.1306	.1312	.1317
10	.1123	.1136	.1149	.1160	.1170	.1180	.1189	.1197	.1205	.1212
11	0.1004	0.1020	0.1035	0.1049	0.1062	0.1073	0.1085	0.1095	0.1105	0.1113
12	.0891	.0909	.0927	.0943	.0959	.0972	.0986	.0998	.1010	.1020
13	.0782	.0804	.0824	.0842	.0860	.0876	.0892	.0906	.0919	.0932
14	.0677	.0701	.0724	.0745	.0765	.0783	.0801	.0817	.0832	.0846
15	.0575	.0602	.0628	.0651	.0673	.0694	.0713	.0731	.0748	.0764
16	0.0476	0.0506	0.0534	0.0560	0.584	0.0607	0.0628	0.0648	0.0667	0.0685
17	.0379	.0411	.0442	.0471	.0497	.0522	.0546	.0568	.0588	.0608
18	.0283	.0318	.0352	.0383	.0412	.0439	.0465	.0489	.0511	.0532
19	.0188	.0227	.0263	.0296	.0328	.0357	.0385	.0411	.0436	.0459
20	.0094	.0136	.0175	.0211	.0245	.0277	.0307	.0335	.0361	.0386
21	0.0000	0.0045	0.0087	0.0126	0.0163	0.0197	0.0229	0.0259	0.0288	0.0314
22	—	—	.0000	.0042	.0081	.0118	.0153	.0185	.0215	.0244
23	—	—	—	—	.0000	.0039	.0076	.0111	.0143	.0174
24	—	—	—	—	—	—	.0000	.0037	.0071	.0104
25	—	—	—	—	—	—	—	—	.0000	.0035

Source: Shapiro and Wilk (1965:603–604); reproduced by permission of *The Biometrika Trustees.*

TABLE 3.2 10%, 5% AND 1% CRITICAL VALUES FOR THE SHAPIRO-WILK TEST FOR NORMALITY

n	0.10	0.05	0.01
3	0.789	0.767	0.753
4	.792	.748	.687
5	.806	.762	.686
6	0.826	0.788	0.713
7	.838	.803	.730
8	.851	.818	.749
9	.859	.829	.764
10	.869	.842	.781
11	0.876	0.850	0.792
12	.883	.859	.805
13	.889	.866	.814
14	.895	.874	.825
15	.901	.881	.835
16	0.906	0.887	0.844
17	.910	.892	.851
18	.914	.897	.858
19	.917	.901	.863
20	.920	.905	.868
21	0.923	0.908	0.873
22	.926	.911	.878
23	.928	.914	.881
24	.930	.916	.884
25	.931	.918	.888
26	0.933	0.920	0.891
27	.935	.923	.894
28	.936	.924	896
29	.937	.926	.898
30	.939	.927	.900
31	0.940	0.929	0.902
32	.941	.930	.904
33	.942	.931	.906
34	.943	.933	.908
35	.944	.934	.910
36	0.945	0.935	0.912
37	.946	.936	.914
38	.947	.938	.916
39	.948	.939	.917
40	.949	.940	.919
41	0.950	0.941	0.920
42	.951	.942	.922
43	.951	.943	.923
44	.952	.944	.924
45	.953	.945	.926
46	0.953	0.945	0.927
47	.954	.946	.928
48	.954	.947	.929
49	.955	.947	.929
50	.955	.947	.930

Source: Shapiro and Wilk (1965:605); reproduced by permission of *The Biometrika Trustees.*

bution, if their parameters differ, then these differences could be what is being measured (Griffith 1981a, 1981b). For instance, if all X values are normally distributed but $\mu_1 \neq \mu_2 \neq \ldots \neq \mu_n$, a spatial autocorrelation statistic value could be attributable to the nature and degree of autocorrelation in the population, sampling error, and/or these differences in mean. In addition, if $\sigma_1^2 \neq \sigma_2^2 \neq \ldots \neq \sigma_n^2$, then a fourth possible source exists. In order to circumvent these problems, then, one must assume that all areal unit variates X have the same distribution and parameters. Moreover, all X_i values have identical normal distributions, with a mean of μ and a variance of σ^2. Whenever this assumption is not palatable, the randomization perspective should be evoked, unless the parameter differences are systematic and hence can be summarized by a few additional parameters, like in a trend surface model.

An Example: Buffalo Crime Data

In the next several chapters a common empirical example of 1981 Buffalo crime data will be used to illustrate many notions. Figure 3.1 presents the map

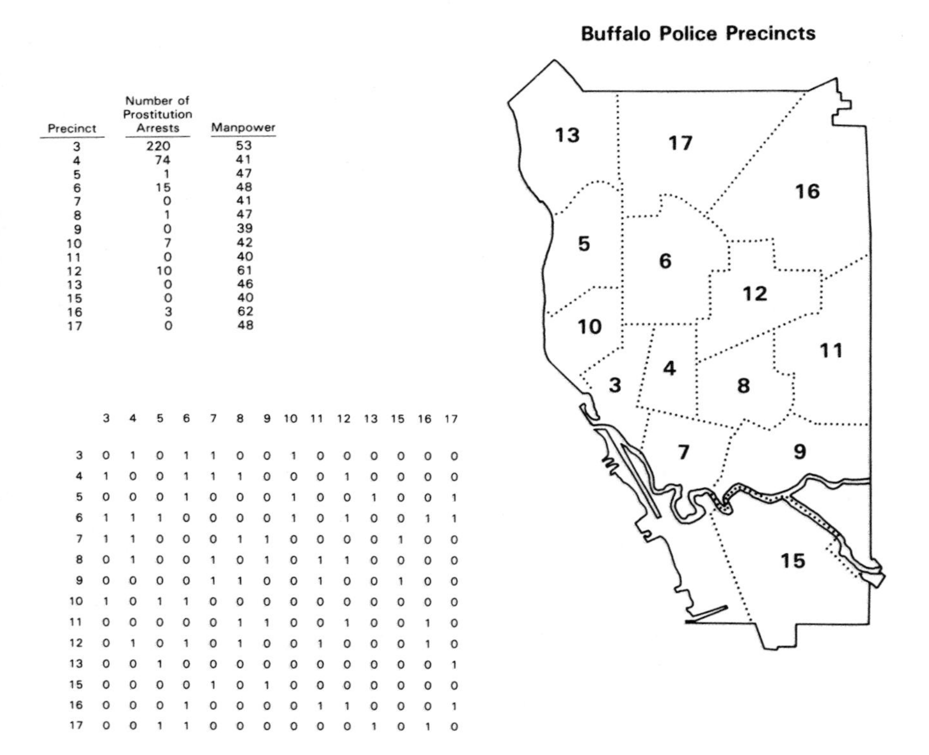

Precinct	Number of Prostitution Arrests	Manpower
3	220	53
4	74	41
5	1	47
6	15	48
7	0	41
8	1	47
9	0	39
10	7	42
11	0	40
12	10	61
13	0	46
15	0	40
16	3	62
17	0	48

	3	4	5	6	7	8	9	10	11	12	13	15	16	17
3	0	1	0	1	1	0	0	1	0	0	0	0	0	0
4	1	0	0	1	1	1	0	0	0	1	0	0	0	0
5	0	0	0	1	0	0	0	1	0	0	1	0	0	1
6	1	1	1	0	0	0	0	1	0	1	0	0	1	1
7	1	1	0	0	0	1	1	0	0	0	0	1	0	0
8	0	1	0	0	1	0	1	0	1	1	0	0	0	0
9	0	0	0	0	1	1	0	0	1	0	0	1	0	0
10	1	0	1	1	0	0	0	0	0	0	0	0	0	0
11	0	0	0	0	0	1	1	0	0	1	0	0	1	0
12	0	1	0	1	0	1	0	0	1	0	0	0	1	0
13	0	0	1	0	0	0	0	0	0	0	0	0	0	1
15	0	0	0	0	1	0	1	0	0	0	0	0	0	0
16	0	0	0	1	0	0	0	0	1	1	0	0	0	1
17	0	0	1	1	0	0	0	0	0	0	1	0	1	0

FIGURE 3.1 1981 Buffalo Police Precincts, and the Geographic Distributions of Police Manpower and Prostitution Arrests

of areal units (police precincts) for this example, together with one type of crime data (prostitution). Under the randomization sampling perspective discussed in this chapter, one has the following frequencies to consider:

Number of Arrests	Frequency of Precincts
220	1
74	1
15	1
10	1
7	1
3	1
1	2
0	6

Accordingly the number of possible geographic distributions of these 14 numbers over the map is $14!/(1!\times 1!\times 1!\times 1!\times 1!\times 1!\times 2!\times 6!) = 60540480$. Figure 3.1, the observed geographic distribution, is just one of these more than sixty-million possibilities. Because there are so many possible map patterns, the subsequent empirical examples will be augmented with simple, contrived numerical examples having a small number of possible map patterns.

The second sampling perspective can be illustrated with the manpower figures in Figure 3.1. If a researcher wanted to conduct indepth interviews of several policemen from each precinct, say two, a sample drawn to ensure complete coverage of the city (a stratified random sample, with the precincts as the stratification), then the total number of samples is

Precinct	Number of Samples
3	C(53,2) = 1378
4	C(41,2) = 820
5	C(47,2) = 1081
6	C(48,2) = 1128
7	C(41,2) = 820
8	C(47,2) = 1081
9	C(39,2) = 741
10	C(42,2) = 861
11	C(40,2) = 780
12	C(61,2) = 1830
13	C(46,2) = 1035
15	C(40,2) = 780
16	C(62,2) = 1891
17	C(48,2) = 1128
	total = 1.92×10^{42}

where C(N,n) denotes the combination of N items (the number of policemen by precinct) taken n (2) at a time. The sampling distribution of, say, the average number of arrests per officer in each precinct would be approximately normally distributed, yielding a sampling distribution in Precinct 3 based upon 1378 values, one in Precinct 4 based upon 820 values, and so forth. The stratified random sample of size 28 that would be drawn then would have approximately a multivariate normal distribution.

4

Spatial Autocorrelation as a Parameter of Statistical Models

As the preceding chapters indicate, conventional statistical analysis of spatial data series can benefit from the explicit treatment of spatial autocorrelation. Several approaches have been taken to this particular problem (Griffith 1986). The most promising results to date have been obtained from the introduction of an additional parameter into conventional statistical models that accounts for the latent spatial autocorrelative structure. This approach has proven to be the best one in time series modeling, too.

The Spatial Autoregressive Model

The statistical problem being addressed revolves around the issue of what complications arise when the null hypothesis, that there is zero spatial autocorrelation in the population, is rejected. Recent literature is replete with arguments and demonstrations that model specification errors, biased parameter estimations, and other problems result from the presence of non-zero spatial autocorrelation. In terms of classical statistics, the assumption of independent observations is violated. This is particularly important since most if not all classical statistics do not seem to be robust with regard to this assumption. Moreover, the statistical geographic sampling problem more closely resembles an experiment in which clusters of grapes, rather than individual balls, are drawn from an urn (Stephan 1934). Modelling endeavors being developed to handle this situation are complex and sophisticated, and this section will serve only to bridge the context of this monograph to these pursuits.

The most popular approach, which is consistent with a position of total ignorance and uncertainty, has been to cast each x_i value as a function of the average of its juxtaposed x_j values. Thus, if each x_j value is multiplied by $w_{ij} = 1/n_i$, where n_i is the number of areal units in juxtaposition with areal unit i, then this average term may be written as

$$\sum_{j=1}^{j=n} w_{ij} x_j .$$

As such, $w_{ij} = c_{ij} / \sum_{j=1}^{j=n} c_{ij}$, where $c_{ij} = 1$ if areal units i and j are adjacent and 0 otherwise. Meanwhile, the nature and degree of this relationship may be captured in a parameter, say ρ, which actually is the autocorrelation parameter. Furthermore, since there could be some type of stochastic error affiliated with each x_i, then the relationship being outlined here entails partitioning each x_i into a systematic, autocorrelation term, and a random error term, say u_i. Therefore, let

$$x_i = \rho \sum_{j=1}^{j=n} w_{ij} x_j + u_i . \tag{4.1}$$

Attaching this equation as a constraint to classical statistical models leads to results that are interpretable within traditional statistical settings. Such interpretations, as well as methods for estimating ρ, will serve as the subjects of this section. One should note that equation (4.1) is known as the "simultaneous" model, but one of many possible models of spatial dependency.

Estimating the Spatial Autocorrelation Parameter ρ

Equation (4.1) may be rewritten as

$$u_i = x_i - \rho \sum_{j=1}^{j=n} w_{ij} x_j , \tag{4.2}$$

and by assumption the u_i values are independently and identically normally distributed. Hence, a maximum likelihood estimate of ρ can be obtained by expressing the joint normal probability density function for n areal units as the following likelihood function:

$$\begin{aligned} L &= (\sigma_u \sqrt{2\pi})^{-n} \exp\{-\sum_{i=1}^{i=n} [(u_i - \mu_u)^2/(2\,\sigma_u^2)]\} \\ &= [\prod_{i=1}^{i=n} (1 - \rho\,\lambda_i)](\sigma_u \sqrt{2\pi})^{-n} \times \\ &\quad \exp\{-\sum_{i=1}^{i=n} [(x_i - \rho \sum_{j=1}^{j=n} w_{ij} x_j) - \mu_u]^2/(2\,\sigma_u^2)\} , \end{aligned} \tag{4.3}$$

where λ_i are the eigenvalues of the stochastic version of the connectivity matrix **C**.

The term $[\prod_{i=1}^{i=n} (1 - \rho\,\lambda_i)]$ is the Jacobian of the transformation to the unautocorrelated domain u_i from the autocorrelated domain x_i. A discussion of eigenvalues and Jacobians is beyond the scope of this monograph; the interested student should consult an intermediate calculus text and a linear algebra text if these concepts are unfamiliar. The important feature of the Jacobian to note here is that it acts as a constraint on the estimation of ρ. Its presence also signifies that classical statistical estimation procedures can not be used to calculate $\hat{\rho}$, indicating that spatial autocorrelation introduces complications into statistical analyses that lie dormant in conventional statistical work. The likelihood function has three parameters, namely μ_u, σ_u^2 and ρ. These three parameters mean that eight estimating equations can be derived from the likelihood function, depending upon which parameter values are known. These possibilities are outlined elsewhere (Griffith 1981a); this monograph is concerned only with the most probable of these eight cases, namely all three parameter values are unknown. Adhering to mathematical statistical theory procedures, these parameters can be estimated by minimizing this likelihood function in its logarithmic form (see Appendix B).

Sampling Distribution Characteristics of $\hat{\rho}$

The optimization of equation (4.3) remains a non-linear problem even if $\ell n(L)$ is optimized. Hence evaluation of the statistical properties of the parameter estimates $\hat{\mu}$, $\hat{\sigma}^2$ and $\hat{\rho}$ is more difficult than in conventional settings. Generalized least squares types of estimates for μ and σ^2 area easily determined. Asymptotic estimates of the standard errors of all three of these parameter estimates are also available. The simplest of these estimates are as follows:

parameter	estimate	asymptotic standard error
μ	$[\sum_{i=1}^{i=n} x_i - \rho \sum_{i=1}^{i=n} \sum_{j=1}^{j=n} (w_{ij} + w_{ji})x_j + \rho^2 \sum_{i=1}^{i=n} \sum_{j=1}^{j=n} \sum_{k=1}^{k=n} w_{ik}w_{kj}x_j]/[n(1-\rho)^2]$	$\sigma^2/[n(1-\rho)^2]$
σ^2	$[\sum_{i=1}^{i=n} (x_i - \bar{x})^2 - \rho \sum_{i=1}^{i=n} \sum_{j=1}^{j=n} (w_{ij} + w_{ji})(x_i - \bar{x})(x_j - \bar{x}) + \rho^2 \sum_{i=1}^{i=n} \sum_{j=1}^{j=n} \sum_{k=1}^{k=n} w_{ik}w_{kj}(x_i - \bar{x})(x_j - \bar{x})]/n$	no closed form
ρ	no closed form	no closed form

The inability to determine closed form solutions in some of these cases reiterates the notion that difficulties lying dormant in conventional statistical analysis emerge during the analysis of spatial data series. These sorts of problems have encouraged spatial analysts to do nothing more than test for the presence of non-zero spatial autocorrelation.

5

Measuring Spatial Autocorrelation

Before surveying popular ways of measuring spatial autocorrelation, one needs to recall that classical correlation has different coefficients for the different kinds of measurement scales. Spatial autocorrelation indices are measurement scale specific, too.

The Join Count Statistic

If data are measured with a nominal scale, then either an areal unit has or does not have the characteristic in question. Let an areal unit be coded with a one if the characteristic in question is present, and a zero otherwise. Then n_1 areal units are coded unity, and n_2 areal units are coded zero, where $n_1 + n_2 = n$. Hence $P(1) = n_1/n$ and $P(0) = n_2/n$.

Suppose that WW denotes the number of juxtaposed areal units for which both units in a pair have been coded with unity. Further, suppose that BB denotes the number of juxtaposed areal units for which both units in a pair have been coded with zero. Finally, suppose that BW denotes the number of juxtaposed areal units for which one unit in a pair has been coded with unity while the other has been coded with zero. Clearly the sum BB + BW + WW equals all connections in the planar partitioning, or half the number of ones that appear in matrix **C**, to be denoted here by J = BB + BW + WW.

Using an underlying *random sampling* perspective, the total number of possible map patterns is 2^n. The expected value of BB in this case is Jn_1^2/n^2, the expected value of BW is $2Jn_1n_2/n^2$, and the expected value of WW is Jn_2^2/n^2. These values are derived from the binomial model. Obviously $P(1) + P(0) = n_1/n + n_2/n = 1$. Since dyads of areal units are being selected, then the model is

$$(n_1/n + n_2/n)^2 = n_1^2/n^2 + 2\,n_1\,n_2/n^2 + n_2^2/n^2.$$

The first term of this expression indicates the probability of randomly selecting a pair of areal units such that each is coded with a one, given that this coding was done in a haphazard, independent fashion. This term is both a probability and a percentage, since $1^2 = 1$. Multiplying this probability by the total number of pairs, J, yields the expected value expression. Similarly, the second term of this expression indicates the probability of randomly selecting a pair of areal units such that one is coded with unity and the other is coded with a zero. The 2 appears because there are two ways of making this selection; either the unity coded areal unit is drawn first, or it is drawn second. Once again, multiplying this probability by J renders the expected value. The third expression refers to a situation analogous to that for the first term, except now both areal units are coded with a one.

Using a *randomization perspective,* the total number of possible map patterns is $n!/(n_1!n_2!)$. The probability of selecting a unity coded areal unit on the first draw is still n_1/n. But permutations refer to sampling without replacement. Accordingly, if a unity coded areal unit already has been selected, then there are only $(n_1 - 1)$ remaining such units, and $(n - 1)$ remaining areal units altogether. The conditional probability for drawing a second unity coded areal unit, then, is $(n_1 - 1)/(n - 1)$. Hence, the joint probability for random selections becomes $n_1(n_1 - 1)/[n(n - 1)]$; multiplying this percentage by J yields the corresponding expected value.

If two differently coded areal units are drawn, then probabilities for the selection sequence are either n_1/n followed by $n_2/(n - 1)$, or n_2/n followed by $n_1/(n - 1)$. Regardless, the product of these pairs of probabilities is $n_1n_2/[n(n - 1)]$, which again needs to be multiplied by 2 because of the two distinct ways this type of sample can be drawn. The expected value for this case, then, is $2Jn_1n_2/[n(n - 1)]$.

Paralleling the procedure for selecting two unity coded areal units, if two zero coded areal units are selected, then the expected value is given by $Jn_2(n_2 - 1)/[n(n - 1)]$.

As similarly coded areal units tend to be in juxtaposition with one another, $BW \rightarrow 0$ and either BB increases, WW increases, or both, depending upon the magnitudes of n_1 and n_2. On the other hand, as dissimilarly coded areal units tend to be in juxtaposition with one another, then $BB \rightarrow 0$ and $WW \rightarrow 0$, and $BW \rightarrow J$.

Consider a planar surface that has been partitioned into four areal units, such as the one appearing in Figure 5.1. Let $n_1/n = n_2/n \Rightarrow n_1 = n_2 = 2$. In this case $J = 4$, since the corresponding connectivity matrix **C** is

$$\begin{matrix} 0 & 1 & 1 & 0 \\ 1 & 0 & 0 & 1 \\ 1 & 0 & 0 & 1 \\ 0 & 1 & 1 & 0 \end{matrix}$$

If the random sampling assumption holds, then the sample space consists of spatial distributions A thru P in Figure 5.1. These distributions have the following counts:

spatial distribution	BB	BW	WW
A	0	0	4
B	0	2	2
C	0	2	2
D	0	2	2
E	0	2	2
F	1	2	1
G	1	2	1
H	1	2	1
I	1	2	1
J	0	4	0
K	0	4	0
L	2	2	0
M	2	2	0
N	2	2	0
O	2	2	0
P	4	0	0

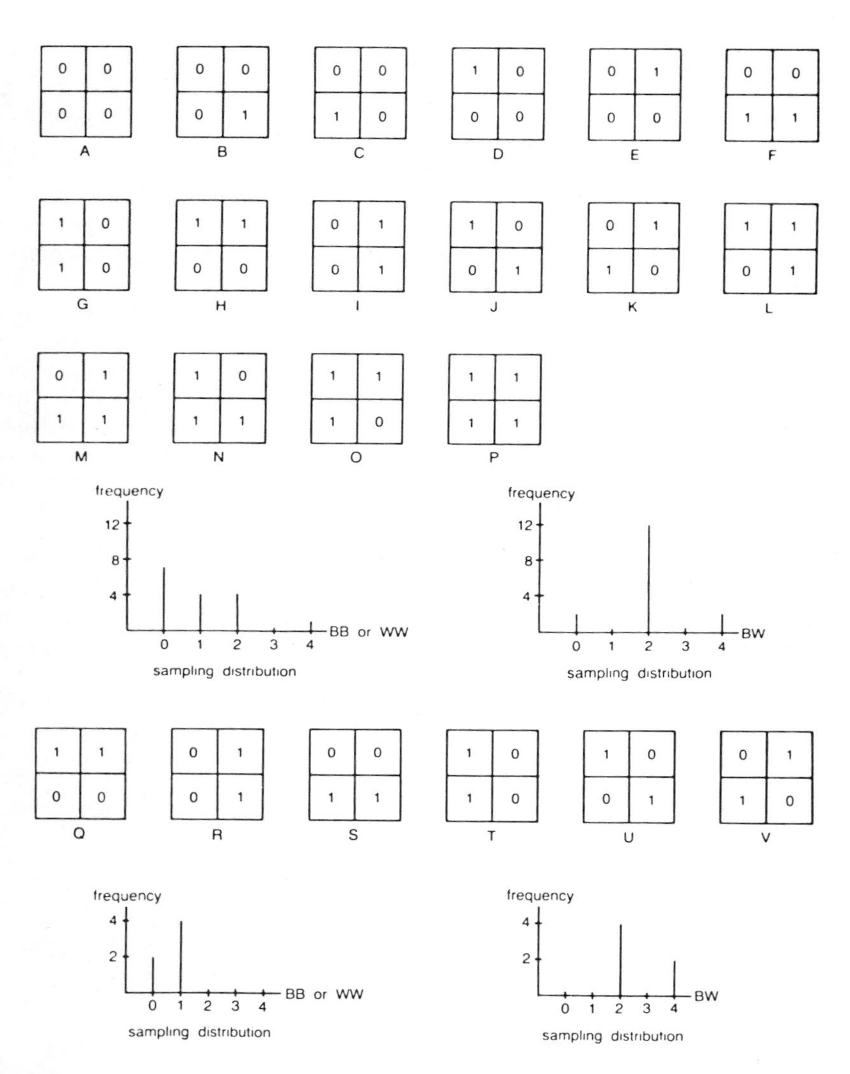

FIGURE 5.1 Sampling Distribution of the Join Count Statistics for a Simple Geographic Landscape

The expected value $\mu_{BB} = 16/16 = 4(1/2)(1/2)$. The expected value $\mu_{BW} = 31/16 = (2)(4)(1/2)(1/2)$. And, the expected value $\mu_{WW} = 16/16 = 4(1/2)(1/2)$.

Meanwhile, if the randomization assumption holds, then the sample space consists of spatial distributions Q thru V in Figure 5.1. These distributions have the following counts:

spatial distribution	BB	BW	WW
Q	1	2	1
R	1	2	1
S	1	2	1
T	1	2	1
U	0	4	0
V	0	4	0

The expected value $\mu_{BB} = 4/6 = 4(2/4)(1/3)$. The expected value $\mu_{BW} = 16/6 = (2)(4)(2/4)(2/3)$. And, the expected value $\mu_{WW} = 4/6 = 4(2/4)(1/3)$.

The most reliable results for this autocorrelation measure occur when $n > 20$, and preferably $n > 30$, and neither $n_1/n < .2$ nor $n_2/n < .2$ (Cliff and Ord 1981a). The best results are obtained when $n_1 = n_2$ (Silk 1979). If $n_1/n > n_2/n_1$ then the test statistic should be based on WW; if $n_2/n > n_1/n$, then the test statistic should be based on BB. Finally, BW reduces to the classical Runs statistic, which was mentioned earlier, when a linear configuration of areal units is being analyzed (Cliff and Ord 1981a:11).

The Buffalo Crime Data Example: Part I

Returning to Figure 3.1, one could ask whether or not the presence/absence of prostitution arrests was spatially autocorrelated (Figure 5.2). Using the

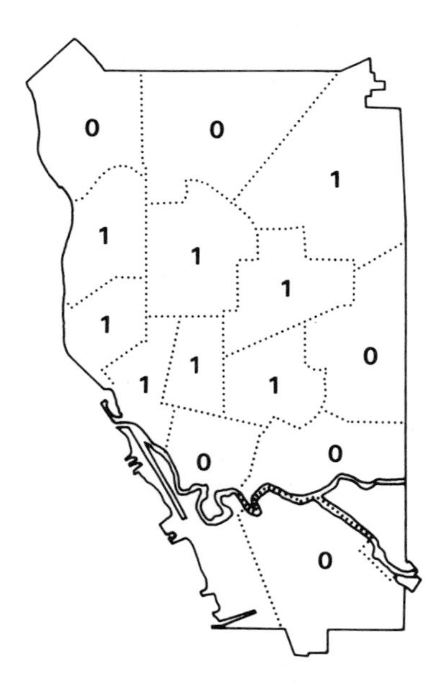

FIGURE 5.2 Geographic Distribution of the Presence/Absence of 1981 Prostitution Arrests by Buffalo Police Precinct

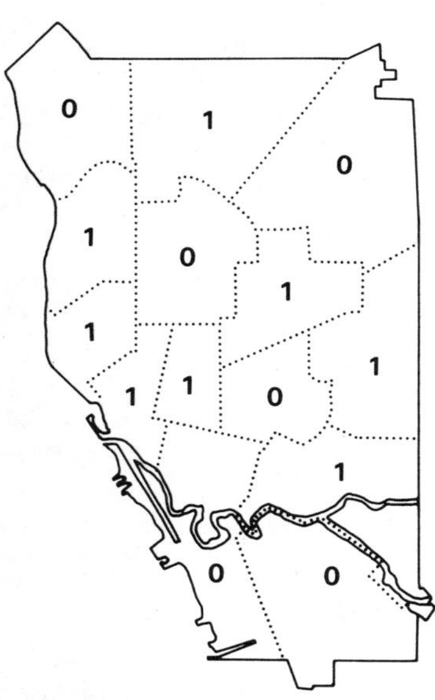

(a) Negative Spatial Autocorrelation

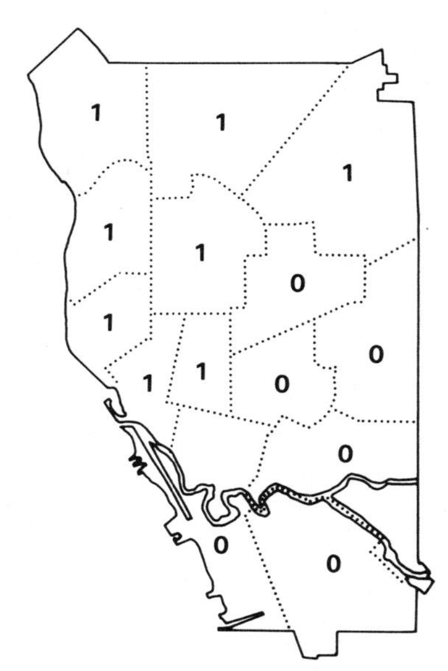

(b) Positive Spatial Autocorrelation

FIGURE 5.3 Map Pattern Extremes for the Possible Join Count Statistic Situation of Figure 5.2, Using Randomization

randomization perspective, the number of possible map patterns is $14!/(8! \times 6!) = 3003$. Using a random sampling perspective, this number becomes $2^{14} = 16384$. The associated join count statistics together with their corresponding population parameters are

Statistic	Observed	Expected Values	
		Random Sampling	Randomization
WW	13	$29(8^2/14^2) = 9.5$	$29[8 \times 7/(14 \times 13)] = 8.9$
BW	11	$2 \times 29(8 \times 6/14^2) = 14.2$	$2 \times 29[8 \times 6/(14 \times 13)] = 15.3$
BB	5	$29(6^2/14^2) = 5.3$	$29[6 \times 5/(14 \times 13)] = 4.8$

The total sum of c_{ij} values is 58 (Figure 3.1), and so $J = 58/2 = 29$.

Figure 5.3 presents two maps of extreme patterns. Using the randomization sampling perspective, in Figure 5.3a the set of zeroes and ones has been arranged in an attempt to maximize negative spatial autocorrelation, whereas in Figure 5.3b the set of zeroes and ones has been arranged in an attempt to maximize positive spatial autocorrelation. The resulting join count statistics are

Statistic	Figure 5.3a	Figure 5.3b
WW	7	13
BW	19	7
BB	3	9

These results indicate a weak degree of positive spatial autocorrelation, since the observed WW value is closest to that for Figure 5.3b, while the observed BW and BB values are closest to the expected ones under the null hypothesis of a random arrangement of zeroes and ones over the map.

The remainder of this mongraph will focus on data measured at the interval/ratio scale, because of its more common occurrence. The student who is interested in pursuing nominal data issues should consult Silk (1979), among others.

The Geary Ratio

One index that can be applied to interval/ratio data is the Geary Ratio (Geary 1954). This index is based upon a paired comparison of juxtaposed map values. Its formula is

$$GR = [(n-1) \sum_{i=1}^{i=n} \sum_{j=1}^{j=n} c_{ij} (x_i - x_j)^2]/[2(\sum_{i=1}^{i=n} \sum_{j=1}^{j=n} c_{ij}) \sum_{i=1}^{i=n} (x_i - \bar{x})^2] \qquad (5.1)$$

where $c_{ij} = 1$ if areal units i and j are adjacent, and 0 otherwise.

The expected value of GR is unity. In other words, if all possible permutations of a set of values $\{x_i\}$ over n areal units are enumerated, and GR is calculated for these $n!/[\prod_{j=1}^{j=k} (n_j!)]$ distributions, then the average of these GRs would be one (see Appendix C). If random samples of size m_i are selected from a normal distribution, and GR is calculated for all $\prod_{i=1}^{i=n} (N_i)^{m_i}$ spatial distributions that are enumerated, then the average of these GRs would be unity (see Appendix D). Meanwhile, as similar x_i values tend to be in juxtaposition with one another, then $(x_i - x_j)^2 \rightarrow 0$, and hence $GR \rightarrow 0$. On the other hand, as dissimilar x_i values tend to be in juxtaposition with one another, then $GR \rightarrow 2$, although 2 is not a strict upper limit (see Amrhein, Guevara and Griffith 1983).

Consider the set of values {1,2,2,3,3,3,4,4,5} together with the planar partitioning presented in Figure 5.4. One could allocate these 9 values to this configuration of areal units 9!/(1!2!3!2!1!) or 15120 different ways. Letting the allocation rule that is used be one in which similar values become juxtaposed, the spatial distribution that appears in Figure 5.4 could result. The arithmetic mean in this case is

$$\bar{x} = (1 + 2 + 3 + 2 + 3 + 4 + 3 + 4 + 5)/9 = 3,$$

and the variance term is

$$\sum_{i=1}^{i=n} (x_i - \bar{x})^2 = [(1-3)^2 + (2-3)^2 + (3-3)^2 + (2-3)^2 + (3-3)^2 + (4-3)^2 + (3-3)^2 + (4-3)^2 + (5-3)^2] = 12.$$

The numerator term of GR that refers to paired comparisons, as well as the denominator term of the first fraction, relate to the connectivity matrix **C** that was described in a preceding section. To briefly recapitulate, the rows and columns of matrix **C** are labelled with the same sequence of areal units, and

A	B	C
D	E	F
G	H	I

$\overline{X} = 3$

$\Sigma(X_i - \overline{X})^2 = 12$

$n = 9$

$\sum_{i=1}^{i=n} \sum_{j=1}^{j=n} C_{ij} = 24$

	A	B	C	D	E	F	G	H	I
A	0	1	0	1	0	0	0	0	0
B	1	0	1	0	1	0	0	0	0
C	0	1	0	0	0	1	0	0	0
D	1	0	0	0	1	0	1	0	0
E	0	1	0	1	0	1	0	1	0
F	0	0	1	0	1	0	0	0	1
G	0	0	0	1	0	0	0	1	0
H	0	0	0	0	1	0	1	0	1
I	0	0	0	0	0	1	0	1	0

1	2	3
2	3	4
3	4	5

(a) Similar Values Clustering

3	4	3
4	3	2
1	2	5

(b) Random Pattern

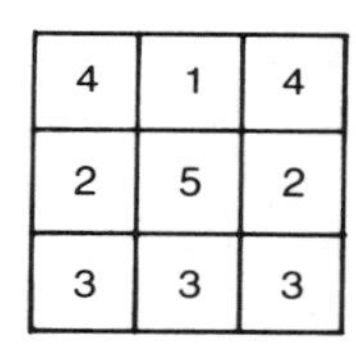

(c) Dissimilar Values Clustering

FIGURE 5.4 Map Pattern Extremes for the Possible Geary Ratio Index Using Randomization

each row-column intersection, denoted c_{ij}, is coded with unity if areal units i and j are adjacent and is coded with zero if these areal units are not contiguous. Matrix **C** for the configuration of 9 areal units presented also appears in Figure 5.4. The denominator term $\sum_{i=1}^{i=n} \sum_{j=1}^{j=n} c_{ij}$ of equation (5.1) states first that all entries in each row are summed, and second that these row sums are added together. Moreover,

$$\text{Step 1: } s_i = c_{i1} + c_{i2} + \ldots + c_{in},$$
$$\text{Step 2: } t_1 = s_1 + s_2 + \ldots + s_n.$$

The terms added together in Step 1 are either zero or one hence s_i equals the number of ones in row i. Further, the terms added together in Step 2 mean that $\sum_{i=1}^{i=n} \sum_{j=1}^{j=n} c_{ij}$ equals the number of ones in matrix **C.** For the example in Figure 5.4 this sum is equal to 24.

The numerator term $\sum_{i=1}^{i=n} \sum_{j=1}^{j=n} c_{ij} (x_i - x_j)^2$ of equation (5.1) states first that the squared difference is calculated between each areal unit value and all areal unit values individually; second, that each squared difference is pre-multiplied by the corresponding connectivity matrix value; third, that these products are summed for each areal unit; and fourth, that the areal unit sums are added together. Moreover,

$$\text{Step 1: } p_i = c_{i1}(x_i - x_1)^2 + c_{i2}(x_i - x_2)^2 + \ldots + c_{in}(x_i - x_n)^2,$$
$$\text{Step 2: } t_2 = p_1 + p_2 + \ldots + p_n.$$

Since c_{ij} is either zero or one, and zero times any number is zero, the terms in Step 1 for which $c_{ij} = 0$ can be ignored. In other words, the only paired comparisons that need to be considered are those of areal units i and j for which a one appears in cell c_{ij} of matrix **C.** Thus, rather than 81 terms to be summed, for the example appearing in Figure 2, only 24 terms need to be summed [in general, rather than n^2 terms, at most only $6(n-2)$ terms will need to be dealt with]. Furthermore, since only the c_{ij} terms are considered, and unity times any number equals that number, these coefficients can be ignored when making the calculation.

Now, regardless of what set of values is distributed over a set of areal units, if the configuration is fixed, then both n and $\sum_{i=1}^{i=n}\sum_{j=1}^{j=n} c_{ij}$ are fixed. If the set of values is fixed, too, and permutations of it are being considered, then since addition is commutative, both $\bar{x}$ and $\sum_{i=1}^{i=n}(x_i - \bar{x})$ are fixed. Hence, for this latter case only the expression $\sum_{i=1}^{i=n}\sum_{j=1}^{j=n} c_{ij}(x_i - x_j)^2$ will change as values are permuted over the partitioned surface. The values for this expression for Figures 5.4a, 5.4b and 5.4c are as follows:

Row-Column Areal Units with $c_{ij} = 1$	Figure 5.4a	Figure 5.4b	Figure 5.4c
A-B	$(1-2)^2$	$(3-4)^2$	$(4-1)^2$
A-D	$(1-2)^2$	$(3-4)^2$	$(4-2)^2$
B-A	$(2-1)^2$	$(4-3)^2$	$(1-4)^2$
B-C	$(2-3)^2$	$(4-3)^2$	$(1-4)^2$
B-E	$(2-3)^2$	$(4-3)^2$	$(1-5)^2$
C-B	$(3-2)^2$	$(3-4)^2$	$(4-1)^2$
C-F	$(3-4)^2$	$(3-2)^2$	$(4-2)^2$
D-A	$(2-1)^2$	$(4-3)^2$	$(2-4)^2$
D-E	$(2-3)^2$	$(4-3)^2$	$(2-5)^2$
D-G	$(2-3)^2$	$(4-1)^2$	$(2-3)^2$
E-B	$(3-2)^2$	$(3-4)^2$	$(5-1)^2$
E-D	$(3-2)^2$	$(3-4)^2$	$(5-2)^2$
E-F	$(3-4)^2$	$(3-2)^2$	$(5-2)^2$
E-H	$(3-4)^2$	$(3-2)^2$	$(5-3)^2$
F-C	$(4-3)^2$	$(2-3)^2$	$(2-4)^2$
F-E	$(4-3)^2$	$(2-3)^2$	$(2-5)^2$
F-I	$(4-5)^2$	$(2-5)^2$	$(2-3)^2$
G-D	$(3-2)^2$	$(1-4)^2$	$(3-2)^2$
G-H	$(3-4)^2$	$(1-2)^2$	$(3-3)^2$
H-E	$(4-3)^2$	$(2-3)^2$	$(3-5)^2$
H-G	$(4-3)^2$	$(2-1)^2$	$(3-3)^2$
H-I	$(4-5)^2$	$(2-5)^2$	$(3-3)^2$
I-F	$(5-4)^2$	$(5-2)^2$	$(3-2)^2$
I-H	$(5-4)^2$	$(5-2)^2$	$(3-2)^2$

The respective GR values for these three examples are

$$GR_{5.4a} = \{(9-1)/[2(24)]\}(24/12) = 1/3\ ,$$
$$GR_{5.4b} = \{(9-1)/[2(24)]\}(72/12) = 1\ , \text{ and}$$
$$GR_{5.4c} = \{(9-1)/[2(24)]\}(132/12) = 11/6\ .$$

As expected, Figure 5.4a has similar values clustering, and hence has a GR value close to zero. Figure 5.4b has no pattern, and hence has a GR value of unity. And, Figure 5.4c has dissimilar values clustering, and hence has a GR value close to 2. Consequently, the Geary Ratio is inversely related to the similarity of juxtaposed map values.

The Buffalo Crime Data Example: Part II

In order to illustrate both positive and negative spatial autocorrelation, consider the 1981 distribution of population and of number of arsons by precinct (see Figure 5.5). Here the calculations become:

Row-Column Precinct with $c_{ij} = 1$	Population (in 1000s)	Arsons
3-4, 4-3	93.30	100
3-6, 6-3	1040.39	400
3-7, 7-3	4.63	3481
3-10, 10-3	415.43	1600
4-6, 6-4	510.58	100
4-7, 7-4	56.37	2401
4-8, 8-4	84.36	529
4-12, 12-4	347.86	3249
5-6, 6-5	5.72	225
5-10, 10-5	89.89	25
5-13, 13-5	64.90	324
5-17, 17-5	28.14	576
6-10, 10-6	140.97	400
6-12, 12-6	15.56	2209
6-16, 16-6	429.36	1024
6-17, 17-6	8.49	81
7-8, 8-7	278.66	676
7-9, 9-7	360.09	784
7-15, 15-7	583.32	1521
8-9, 9-8	5.21	4
8-11, 11-8	11.28	529
8-12, 12-8	89.61	1156
9-11, 11-9	1.16	441
9-15, 15-9	26.79	121
11-12, 12-11	37.31	3249
11-16, 16-11	947.04	1764
12-16, 16-12	608.41	225
13-17, 17-13	178.52	36
16-17, 17-16	317.12	1681

The corresponding Geary Ratio statistics are

$$GR_{population} = (13/2 \times 58)(13560881152/2389775104) = 0.6359\ ,$$
$$GR_{arsons} = (13/2 \times 58)(57822/5242.93) = 1.2360\ .$$

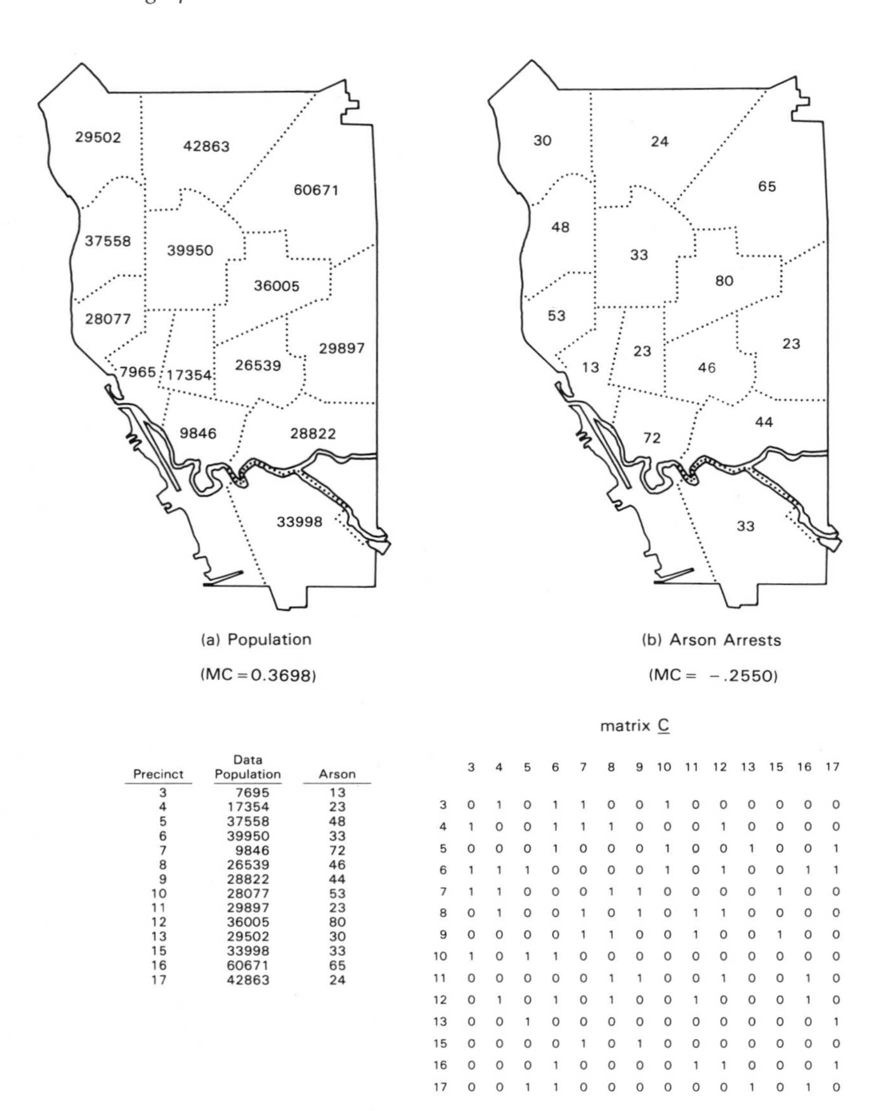

Precinct	Data Population	Arson
3	7695	13
4	17354	23
5	37558	48
6	39950	33
7	9846	72
8	26539	46
9	28822	44
10	28077	53
11	29897	23
12	36005	80
13	29502	30
15	33998	33
16	60671	65
17	42863	24

matrix $\underline{C}$

	3	4	5	6	7	8	9	10	11	12	13	15	16	17
3	0	1	0	1	1	0	0	1	0	0	0	0	0	0
4	1	0	0	1	1	1	0	0	0	1	0	0	0	0
5	0	0	0	1	0	0	0	1	0	0	1	0	0	1
6	1	1	1	0	0	0	0	1	0	1	0	0	1	1
7	1	1	0	0	0	1	1	0	0	0	0	1	0	0
8	0	1	0	0	1	0	1	0	1	1	0	0	0	0
9	0	0	0	0	1	1	0	0	1	0	0	1	0	0
10	1	0	1	1	0	0	0	0	0	0	0	0	0	0
11	0	0	0	0	0	1	1	0	0	1	0	0	1	0
12	0	1	0	1	0	1	0	0	1	0	0	0	1	0
13	0	0	1	0	0	0	0	0	0	0	0	0	0	1
15	0	0	0	0	1	0	1	0	0	0	0	0	0	0
16	0	0	0	1	0	0	0	0	1	1	0	0	0	1
17	0	0	1	1	0	0	0	0	0	0	1	0	1	0

FIGURE 5.5 Geographic Distribution of 1981 Population and Arsons by Buffalo Police Precinct

These statistics suggest that the geographic distribution of population is positively spatially autocorrelated, meaning similar population values tend to cluster together on the map, and that the geographic distribution of arsons is negatively spatially autocorrelated, meaning dissimilar numbers of arsons tend to cluster together. The nature of this second case is somewhat intuitively obvious, since a visual inspection of the map pattern of arsons shows that each of the two largest values (72 and 80) is surrounded by relatively low values.

The Moran Coefficient

Another index that can be applied to interval/ratio data is the Moran Coefficient (Moran 1948). This index is based upon the covariation of juxtaposed map values. Its formula is:

$$MC = n \sum_{i=1}^{i=n} \sum_{j=1}^{j=n} c_{ij}(x_i - \bar{x})(x_j - \bar{x}) / [\sum_{i=1}^{i=n} \sum_{j=1}^{j=n} c_{ij} \sum_{i=1}^{i=n} (x_i - \bar{x})^2] \quad (5.2)$$

The expected value of MC is $-1/(n-1)$ (see Appendix E). Meanwhile, similar to a classical correlation coefficient, as similar x_i values tend to be in juxtaposition with one another, then $MC \rightarrow 1$. Moreover, recalling the foregoing discussion of an ordering of x_i values, if x_i as well as x_j are from the first part of this sequence, then both $(x_i - \bar{x})$ and $(x_j - \bar{x})$ are positive, meaning their product is positive. However, if x_i as well as x_j are from the last part of the sequence, then both $(x_i - \bar{x})$ and $(x_j - \bar{x})$ are negative, meaning their product still is positive. If either x_i or x_j equals $\bar{x}$, then the corresponding product is zero. Therefore, the numerator of equation (5.2) must be positive. On the other hand, as dissimilar x_i values tend to be in juxtaposition with one another, $MC \rightarrow -1$. Now if x_i is from the first part of the sequence of x_i values, x_j would tend to be from the last part of the sequence, or vice versa. Hence one of the terms $(x_i - \bar{x})$ and $(x_j - \bar{x})$ is negative while the other is positive, meaning their product would be negative. Accordingly, the numerator of equation (5.2) must be negative. And, for a random relation, as $n \rightarrow \infty$, $-1/(n-1) \rightarrow 0$.

Returning to the example presented in Figure 5.4, again only the numerator of the second fraction varies if x_i values are permuted over a given partitioned surface. In this instance, though, the values for the expression:

$$\sum_{i=1}^{i=n} \sum_{j=1}^{j=n} c_{ij}(x_i - \bar{x})(x_j - \bar{x})$$

for Figures 5.4a, 5.4b, and 5.4c are as follows:

Row-Column Areal Units with $c_{ij} = 1$	**Figure 5.4a**	**Figure 5.4b**	**Figure 5.4c**
A-B	(1-3)(2-3)	(3-3)(4-3)	(4-3)(1-3)
A-D	(1-3)(2-3)	(3-3)(4-3)	(4-3)(2-3)
B-A	(2-3)(1-3)	(4-3)(3-3)	(1-3)(4-3)
B-C	(2-3)(3-3)	(4-3)(3-3)	(1-3)(4-3)
B-E	(2-3)(3-3)	(4-3)(3-3)	(1-3)(5-3)
C-B	(3-3)(2-3)	(3-3)(4-3)	(4-3)(1-3)
C-F	(3-3)(4-3)	(3-3)(2-3)	(4-3)(2-3)
D-A	(2-3)(1-3)	(4-3)(3-3)	(2-3)(4-3)
D-E	(2-3)(3-3)	(4-3)(3-3)	(2-3)(5-3)
D-G	(2-3)(3-3)	(4-3)(1-3)	(2-3)(3-3)
E-B	(3-3)(2-3)	(3-3)(4-3)	(5-3)(1-3)
E-D	(3-3)(2-3)	(3-3)(4-3)	(5-3)(2-3)
E-F	(3-3)(4-3)	(3-3)(2-3)	(5-3)(2-3)
E-H	(3-3)(4-3)	(3-3)(2-3)	(5-3)(3-3)
F-C	(4-3)(3-3)	(2-3)(3-3)	(2-3)(4-3)
F-E	(4-3)(3-3)	(2-3)(3-3)	(2-3)(5-3)
F-I	(4-3)(5-3)	(2-3)(5-3)	(2-3)(3-3)

G-D	(3-3)(2-3)	(1-3)(4-3)	(3-3)(2-3)
G-H	(3-3)(4-3)	(1-3)(2-3)	(3-3)(3-3)
H-E	(4-3)(3-3)	(2-3)(3-3)	(3-3)(5-3)
H-G	(4-3)(3-3)	(2-3)(1-3)	(3-3)(3-3)
H-I	(4-3)(5-3)	(2-3)(5-3)	(3-3)(3-3)
I-F	(5-3)(4-3)	(5-3)(2-3)	(3-3)(2-3)
I-H	(5-3)(4-3)	(5-3)(2-3)	(3-3)(3-3)

The respective MC values for these three examples are:

$$MC_{5.4a} = (9/24)(16/12) = 1/2\,,$$
$$MC_{5.4b} = (9/24)(-8/12) = -1/4\,, \text{ and}$$
$$MC_{5.4c} = (9/24)(-28/12) = -7/8\,.$$

As expected, once again, Figure 5.4a has similar values clustering, and hence has an MC values close to 1. Figure 5.4b has no pattern, and hence has an MC value close to $-1/8$. And, Figure 5.4c has dissimilar values clustering, and hence has an MC value close to -1. Consequently, the Moran Coefficient is directly related to the similarity of juxtaposed map values.

The Buffalo Crime Data Example: Part III

Returning to the 1981 geographic distributions of population and arsons by precinct (Figure 5.5), the calculations for the Moran Coefficient become

Row-Column Precinct with $c_{ij} = 1$	Population (in 1000s)	Arsons
3-4, 4-3	304.37	547.58
3-6, 6-3	−213.80	258.29
3-7, 7-3	476.55	−869.92
3-10, 10-3	58.48	−320.28
4-6, 6-4	−123.74	169.01
4-7, 7-4	275.82	−569.21
4-8, 8-4	54.26	− 77.07
4-12, 12-4	− 71.38	−720.64
5-6, 6-5	64.62	− 54.21
5-10, 10-5	− 17.67	67.22
5-13, 13-5	− 7.80	− 72.42
5-17, 17-5	84.81	−108.85
6-10, 10-6	− 23.77	− 98.85
6-12, 12-6	50.14	−339.92
6-16, 16-6	280.10	−206.00
6-17, 17-6	114.08	160.08
7-8, 8-7	84.95	122.43
7-9, 9-7	37.51	62.29
7-15, 15-7	− 70.05	−268.50
8-9, 9-8	7.38	8.43
8-11, 11-8	2.98	− 77.07
8-12, 12-8	− 21.99	155.01
9-11, 11-9	1.32	− 39.21
9-15, 15-9	− 6.09	− 18.50
11-12, 12-11	− 3.93	−720.64

11-16, 16-11	− 21.93	−436.71
12-16, 16-12	161.58	878.36
13-17, 17-13	− 13.77	213.86
16-17, 17-16	367.62	−413.64

The corresponding Moran Coefficient statistics are:

$$MC_{population} = (14/58)(3661310464/2389775104) = 0.3698 ,$$
$$MC_{arsons} = (14/58)(-5538.13/5242.93) = -0.2550 .$$

These statistics again suggest that the geographic distribution of population is positively spatially autocorrelated, whereas the geographic distribution of number of arsons is negatively spatially autocorrelated.

Shortcut Calculation Methods

The tediousness of doing calculations outlined in the two preceding sections can be reduced substantially in two ways. First, since matrix **C** is symmetric, $c_{ij} = 1$ means $c_{ji} = 1$. Accordingly, if the dyad A-B is in the sum, then the dyad B-A also must be in the sum. Multiplication is commutative, too. So,

$$(x_i - x_j)^2 = (x_j - x_i)^2 \text{ , and}$$
$$(x_i - \bar{x})(x_j - \bar{x}) = (x_j - \bar{x})(x_i - \bar{x}) .$$

Hence the number of calculations can be cut in half by only considering the upper diagonal of matrix *C* when calculating

$$\sum_{i=1}^{i=n} \sum_{j=1}^{j=n} c_{ij} (x_i - x_j)^2 \text{ and } \sum_{i=1}^{i=n} \sum_{j=1}^{j=n} c_{ij} (x_i - \bar{x})(x_j - \bar{x}) ,$$

and then doubling the result. Moreover,

$$\sum_{i=1}^{i=n} \sum_{j=1}^{j=n} c_{ij} (x_i - x_j)^2 = 2 \sum_{i=1}^{i=n} \sum_{j=1}^{j=i} c_{ij} (x_i - x_j)^2 \text{ , and}$$
$$\sum_{i=1}^{i=n} \sum_{j=1}^{j=n} c_{ij} (x_i - \bar{x})(x_j - \bar{x}) = 2 \sum_{i=1}^{i=n} \sum_{j=1}^{j=i} c_{ij} (x_i - \bar{x})(x_j - \bar{x}) .$$

Second, before calculating MC, the mean $\bar{x}$ could be subtracted from each x_i, and then the spatial distribution of $(x_i - \bar{x})$ could be coded on the map. Since the mean of $(x_i - \bar{x})$ is zero, this modification would substantially reduce the calculations involved. Figure 5.6 presents this modification of the spatial distributions appearing in Figure 5.4.

Relationships Between the Moran Coefficient and the Geary Ratio

As was mentioned earlier, GR→0 whereas MC→1 when similar x_i values tend to cluster on a map. In contradistinction, GR→2 whereas MC→−1 when dissimilar x_i values tend to cluster on a map. Because high values of GR measure what low values of MC measure, and vice versa, these two indices are inversely related.

−2	−1	0
−1	0	1
0	1	2

0	1	0
1	0	−1
−2	−1	2

1	−2	1
−1	2	−1
0	0	0

FIGURE 5.6 The Geographic Distributions Appearing in Figure 5.4, Adjusted for Their Means

Simulation experiments that have been conducted to date suggest that this inverse relationship is basically linear in nature. Departures from linearity, though, may be ascribed to the differences in what each of these two indices measures. The index GR deals with paired comparisons, whereas MC deals with covariations. One index can be expressed in terms of the other, by looking at the ratio of these two kind of similarity measures, as follows:

$$GR = \{(n-1) \sum_{i=1}^{i=n} \sum_{j=1}^{j=n} c_{ij} (x_i - x_j)^2/[2n \sum_{i=1}^{i=n} \sum_{j=1}^{j=n} c_{ij} (x_i - \bar{x})(x_j - \bar{x})]\}MC.$$

Because a planar partitioning is held constant across all spatial distributions used to construct a sampling distribution, the term $(n-1)/(2n)$ plays a minor role in relating MC with GR. The ratio of paired comparisons to covariations is not constant, though, as is attested to by results reported in Table 6.1. Consequently, the functional relation linking GR and MC is specific to a spatial distribution. It is so complex, in fact, that given certain areal unit configurations having particular pairs of juxtaposed values, GR and MC could imply conflicting inferences (Griffith 1982).

6

Constructing Sampling Distributions for the Spatial Autocorrelation Indices

To construct a sampling distribution for some statistic, all possible samples need to be enumerated; the statistic in question must be calculated for each sample; and then a frequency distribution needs to be constructed for the set of sample statistics. Since all possible samples will have been considered, a population has been identified. Measures on this population are for parameters, then.

The Randomization Assumption

Consider the arrangement of four areal units in Figure 6.1. Using a randomization perspective, suppose the set of numbers {0,1,1,2} is to be allocated to this configuration of areal units. The total number of spatial distributions that will result is $4!/(2!1!1!) = 12$. These 12 possibilities are listed in Figure 6.2. One of the problems of classical statistics is illustrated by Figure 6.2. If, for instance, the traditional mean, $\bar{x}$, is calculated for each of these spatial distributions. Because addition is commutative, this statistic always is 2. Thus, the sampling distribution of $\bar{x}$, which also appears in Figure 6.2, has a population mean $\mu_{\bar{x}} = (12)(2)/12 = 2$ and a population variance $\sigma_{\bar{x}}^2 = 12(2-2)^2/12 = 0$. Clearly, arrangement information for the set of x_i values {0,1,1,2} is not explicitly summarized in the traditional statistic $\bar{x}$.

This arrangement information is captured by GR and MC, though. Calculations of these statistics for the sample space appearing in Figure 6.2 are presented in Table 6.1. Now,

$$\mu_{GR} = [8(9/8) + 4(3/4)]/12 = 1\text{, and}$$
$$\sigma_{GR}^2 = [8(9/8 - 1)^2 + 4(3/4 - 1)^2]/12 = 1/32\,.$$

Similarly,

$$\mu_{MC} = [4(0) + 8(-1/2)]/12 = -1/3\text{, and}$$
$$\sigma_{MC}^2 = \{4[0 - (-1/3)]^2 + 8[-1/2 - (-1/3)]^2\}/12 = 1/18\,.$$

Obviously, both MC and GR are summarizing arrangement information. Furthermore, as was noted earlier, the expected value of GR is unity, and the expected value of MC is $-1/(n-1)$.

A	B
C	D

FIGURE 6.1 One Partitioning of a Planar Surface into Four Mutually Exclusive and Collectively Exhaustive Areal Units

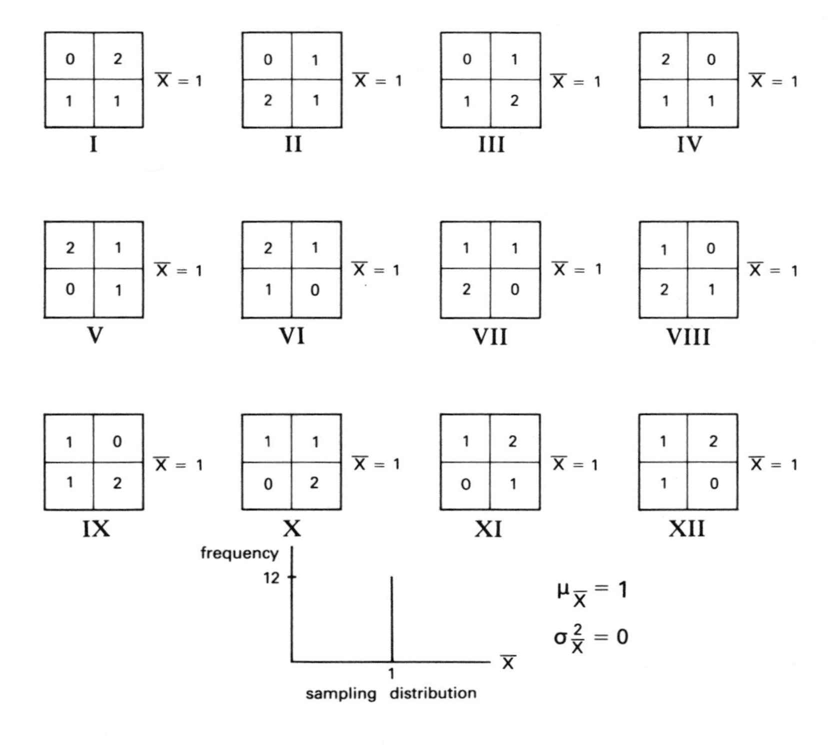

FIGURE 6.2 All Possible Geographic Distributions, Based Upon the Configuration in Figure 6.1 and the Set of Values [0,1,1,2]

Frequency distributions for those calculations appearing in Table 6.1 are presented in Figure 6.3. Because of the simplicity of the example utilized here, these frequency distributions display only two possible statistic values. As was alluded to in Figure 5.2, however, as n increases the number of different GR and MC values increases, and points along the x-axis in the intervals (0,2) and (−1,1), respectively, become feasible, with a concentration around their expected values. One should recall that in Figure 5.4 the three sample spatial distributions, out of 15120 possible distributions, had GR values of 1/3, 1 and 11/6, and MC values of 1/2, −1/4 and −7/8.

TABLE 6.1 GR AND MC STATISTICS FOR THE SAMPLE SPACE APPEARING IN FIGURE 6.2

Configuration	$\sum_{i=1}^{i=n}\sum_{j=1}^{j=n} c_{ij}(x_i - x_j)^2$	GR	$\sum_{i=1}^{i=n}\sum_{j=1}^{j=n} c_{ij}(x_i - \bar{x})(x_j - \bar{x})$	MC
I	6	$\frac{3}{16} \cdot \frac{2 \cdot 6}{2} = \frac{9}{8}$	−1	$\frac{4}{8} \cdot \frac{2(0)}{2} = 0$
II	6	$\frac{9}{8}$	−1	0
III	4	$\frac{3}{16} \cdot \frac{2 \cdot 4}{2} = \frac{3}{4}$	−2	$\frac{4}{8} \cdot \frac{2(-2)}{2} = -1$
IV	6	$\frac{9}{8}$	−1	0
V	6	$\frac{9}{8}$	−1	0
VI	4	$\frac{3}{4}$	−2	−1
VII	6	$\frac{9}{8}$	−1	0
VIII	4	$\frac{3}{4}$	−2	−1
IX	6	$\frac{9}{8}$	−1	0
X	6	$\frac{9}{8}$	−1	0
XI	4	$\frac{3}{4}$	−2	−1
XII	6	$\frac{9}{8}$	−1	0

A Randomization Example for the Buffalo Crime Data

MINITAB can be used to illustrate in a classroom setting the materials outlined in this chapter, especially using its PC compatible microcomputer version. MINITAB was selected here because of its general availability and its highly user-friendly status.

Consider the Buffalo crime data presented in Figure 5.5b, namely arson arrests. Because the values of 23 and 33 repeat in this geographic distribution, there are $14!/[(2!)(2!)] = 21794572800$ possible permutations of these 12 distinct numbers over the 14 police precincts. Clearly even with high-speed computers, constructing the sampling distribution of MC or GR for this many arrangements is prohibitive. But MINITAB has a command called "SAMPLE" that does random sampling without replacement, with order being important. Hence drawing a sample of size 14 from the original 14 numbers is equivalent to permuting these numbers. Accordingly a sample sampling distribution of, say, size 100 could be generated with this particular MINITAB command, in a Monte Carlo fashion.

Suppose the Buffalo arson data are stored in C1 in ascending order of police precinct number:

```
SET C1
13 23 48 33 72 46 44 53 23 80 30 33 65 24
END
```

Further, suppose the connectivity matrix for the 14 precincts (again arranged in ascending order) is stored in matrix M1:

```
READ 14 14 M1
0 1 0 1 1 0 0 1 0 0 0 0 0 0
1 0 0 1 1 1 0 0 0 1 0 0 0 0
0 0 0 1 0 0 0 1 0 0 1 0 0 1
1 1 1 0 0 0 0 1 0 1 0 0 1 1
1 1 0 0 0 1 1 0 0 0 0 1 0 0
0 1 0 0 1 0 1 0 1 1 0 0 0 0
0 0 0 0 1 1 0 0 1 0 0 1 0 0
1 0 1 1 0 0 0 0 0 0 0 0 0 0
0 0 0 0 0 1 1 0 0 1 0 0 1 0
0 1 0 1 0 1 0 0 1 0 0 0 1 0
0 0 1 0 0 0 0 0 0 0 0 0 0 1
0 0 0 0 1 0 1 0 0 0 0 0 0 0
0 0 0 1 0 0 0 0 1 1 0 0 0 1
0 0 1 1 0 0 0 0 0 0 1 0 1 0
END
```

Then the set of MINITAB commands for calculating an MC is:

```
CENTER C1 C2;
LOCATION.
MULT M1 C2 C3
LET K1 = SUM(C2*C3)
LET K2 = SUM(C2**2)
SET C4
14(1)
END
```

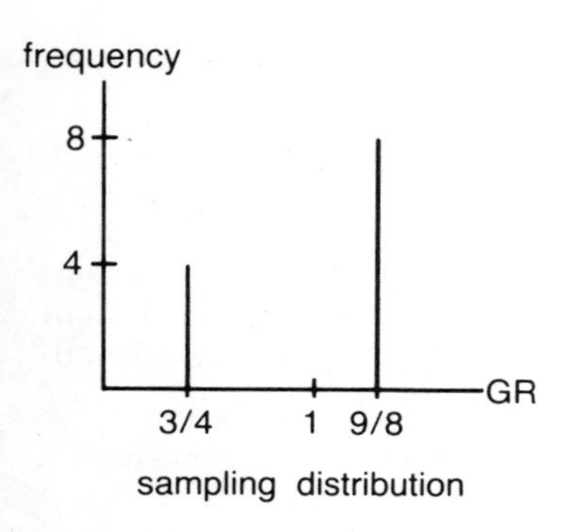

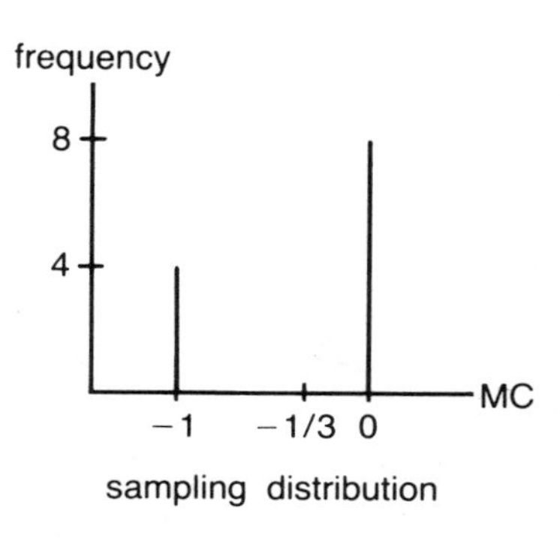

FIGURE 6.3 The Sampling Distribution of GR and MC for Figure 6.2

```
MULT M1 C4 C5
LET K3 = SUM(C5)
LET K4 = (14.0/K3)*(K1/K2)
PRINT K4
```

The command CENTER with subcommand LOCATION calculates $(x_i - \bar{x})$. The commands MULT M1 C2 C3 and LET K1 = SUM(C2*C3) calculate

$$\sum_{i=1}^{i=n} \sum_{j=1}^{j=n} c_{ij}(x_i - \bar{x})(x_j - \bar{x})$$

The command LET K2 = SUM(C2**2) calculates

$$\sum_{i=1}^{i=n} (x_i - \bar{x})^2$$

C4 is a vector of ones; the commands MULT M1 C4 C5 and LET K3 = SUM(C5) calculate

$$\sum_{i=1}^{i=n} \sum_{j=1}^{j=n} c_{ij}$$

Finally, MC is calculated with the command LET K4 = (14.0/K3)*(K1/K2). The command PRINT K4 yields −0.254970.

This set of commands can be stored in a MINITAB subroutine using STORE "SAMPLING", with the additional lines of code

```
LET C100(K10) = K4
LET K10 = K10 + 1
SAMPLE 14 C99 C1
END
```

Then a macro-subroutine can be created that has the commands

```
COPY C1 C99
SAMPLE 14 C99 C1
LET K10 = 1
EXEC "SAMPLING" 100
END
```

The commands HISTOGRAM and DESCRIBE will summarize the resulting sample sampling distribution. For the random samples drawn here (a different set of samples will be drawn each time the routine is executed), the resulting distribution is as follows:

interval midpoint	frequency	interval midpoint	frequency
−.4	1	0	19
−.3	9	.1	13
−.2	24	.2	6
−.1	26	.3	1
		.4	1

$\hat{\mu}_{MC} = -0.0741 \quad (\mu_{MC} = -0.0769)$
$\hat{\sigma}_{MC} = 0.1524$

Using the NSCORES command, the 100 sampled values appear to conform very closely to a normal distribution (the modified Wilk-Shapiro statistic is 0.993).

The Normality Assumption

Let two items exist in each of the four areal units appearing in Figure 6.1, and in each case let these two items take on the respective numerical values of 0 and 2. If a sample of size 2 is drawn in each areal unit, then $(N_i)^{m_i} = 2^2$ for all i. and the total number of possible samples is $\prod_{i=1}^{i=4} 2^2 = 256$. The sample space for each areal unit would be

$$\begin{matrix}(0,0)\ (0,2) \\ (2,0)\ (2,2)\end{matrix} \Rightarrow \begin{matrix}\bar{x} = 0 & \bar{x} = 1 \\ \bar{x} = 1 & \bar{x} = 2\end{matrix}$$

Therefore, if arithmetic means are of interest, then combinations of numbers that can be allocated to the four areal units come from the set {0,1,1,2}.

Since {0,1,1,2} is the set of numbers that is drawn from when constructing a spatial distribution, it should be compared with a normal frequency distribution. Using the Shapiro-Wilk statistic,

$$W = (.6872\,|0-2| + .1677\,|1-1|)^2/[(0-1)^2 + (1-1)^2 + (1-1)^2 + (2-1)^2]$$
$$= .9445.$$

Employing a 10% level of significance, $W > W_{.10,4} = .792$. Consequently, there is no evidence to indicate that the frequency distribution of $\bar{x}$ in the population does not conform to a normal distribution. This inference is not surprising, given the central limit theorem of classical statistics.

Because order is important, in drawing the samples within each areal unit, samples (0,2) and (2,0) are treated as being different, even though they both yield $\bar{x} = 1$. The range of possible sample spatial distributions can be rewritten as a combinatorial problem of selecting sets of the four possible sample compositions. For instance, the sample (0,2) could be selected in all four areal units. In this case, one would select four samples, all of which are the same. Since only one sample type is selected, then the number of possibilities is given by C(4,1), the combinations of 4 units taken 1 at a time. The number of arrangements of this sample composition on the map is given by $4!/4! = 1$, since all four samples are the same. On the other hand, the samples (0,2) and (2,2) might be selected such that one is drawn in three of the areal units, and the other is drawn in one of the areal units. In this case, one would select four samples, in which three are the same and one is different. Since two sample types are selected, then the number of possibilities is given by C(4,2). But, either sample type could constitute the multiple selection. Accordingly, the number of possibilities for this multiple selection is given by C(2,1). The number of arrangements of this sample composition on the map is given by $4!/(3!1!) = 4$, since three samples are the same. Extending these results to all possible sample compositions yields

sample composition	**frequency**	
4 samples the same	C(4,1) × 4!/4!	= 4
3 samples the same, 1 different	C(4,2) × C(2,1) × 4!/(3!1!)	= 48
2 pair of two samples the same	C(4,2) × 4!/(2!2!)	= 36
2 samples the same, 2 different	C(4,3) × C(3,2) × 4!/(2!1!1!)	= 144
4 different samples	C(4,4) × 4!	= 24
	TOTAL	= 256

Regardless that the samples (2,0) and (0,2) are different, they yield the same mean. If $\bar{x}_i$ values are mapped, then the total number of distinct spatial distributions is decreased. For example, the case of four different means becomes infeasible. The other modifications are as follows:

sample composition	**frequency**	
4 samples the same	C(3,1) × 4!/4!	= 3
3 samples the same, 1 different	C(3,2) × C(2,1) × 4!/3!	= 24
2 pair of two samples the same	C(3,2) × 4!/(2!2!)	= 18
2 samples the same, 2 different	C(3,3) × C(3,2) × 4!/2!	= 36
	TOTAL	= 81

The sampling distribution of GR for this modified sample space appears in Figure 6.4. The case of four samples the same must be ignored, since each of the resulting surfaces is constant. The remaining 78 spatial distributions for this illustration constitute the sample space. Parameters of this enumeration of samples are as follows:

$$\mu_{GR} = [(16)(3/4) + (16)(9/11) + (24)(1) + (8)(9/8) + (8)(15/11) + (6)(3/2)]/78 = 1 \text{, and}$$

$$\begin{aligned}\sigma_{GR}^2 &= [16(3/4 - 1)^2 + 16(9/11 - 1)^2 + 24(1 - 1)^2 + \\ &\quad 8(9/8 - 1)^2 + 8(15/11 - 1)^2 + 6(3/2 - 1)^2]/78 \\ &= 1359/25168 \\ &= .054 .\end{aligned}$$

Once again, as anticipated, the expected value of GR is found to be unity. Clearly similar results are found for the MC statistic, where $\mu_{MC} = -1/3$ and $\sigma_{MC}^2 = .096$.

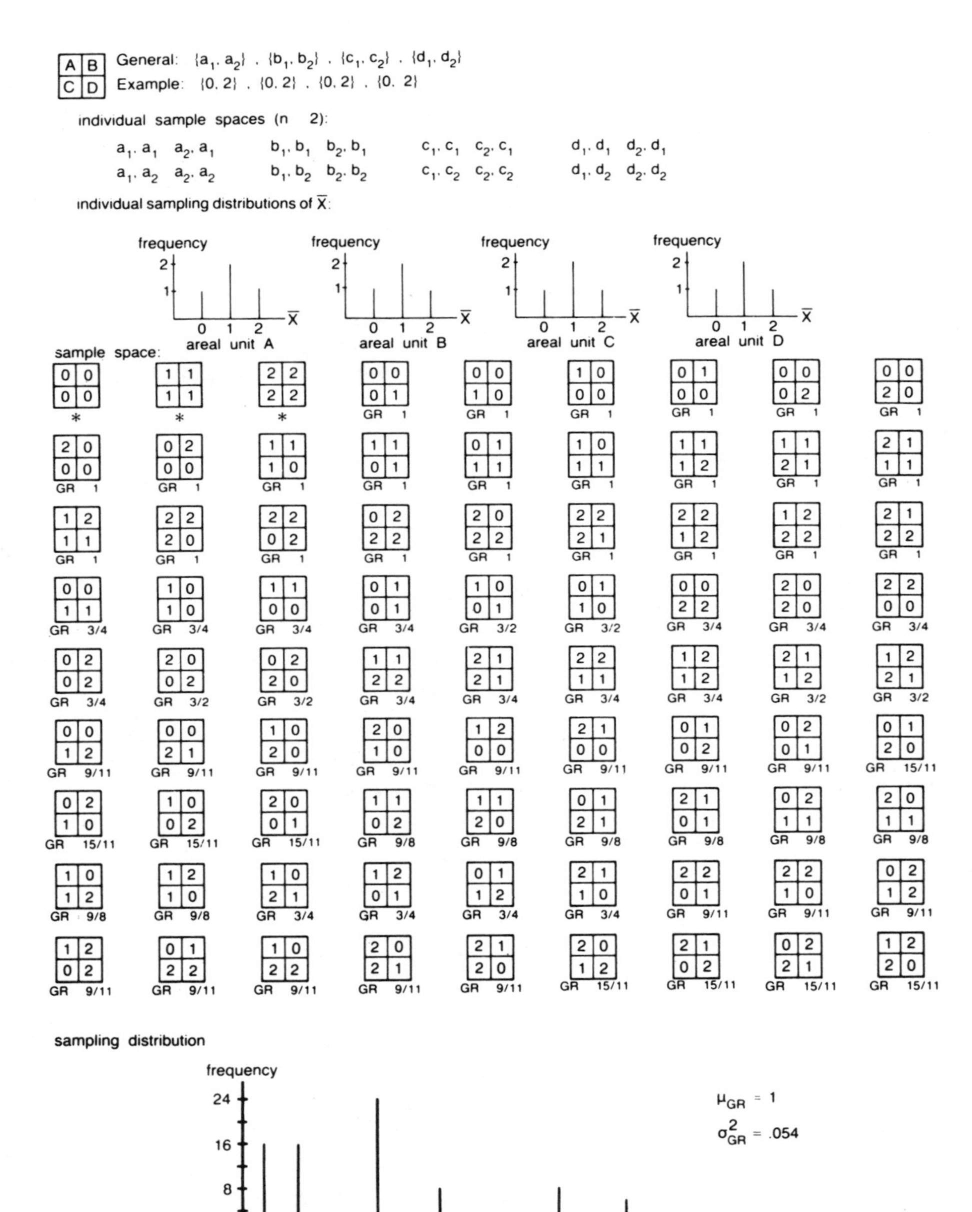

FIGURE 6.4 All Possible Geographic Distributions for a Stratified Random Sample Example

This result relates directly to that obtained for the randomization assumption. When counting the frequency of spatial distributions for a given sampling composition, both a combination term and a permutation term are involved. The combination term counts the number of samples of a specific composition type. The permutation term counts the number of spatial distributions for a given sample composition. The randomization assumption implies that the expected value of GR is unity for each of these sets of spatial distributions. In fact, the example used in the previous section is one of the possible compositions appearing in Figure 6.4, where two samples that are the same and two samples that are different have been drawn. Thus, it is not surprising that the average for all of these sample compositions also is one. In essence, then, the normality assumption coupled with the example described in Figure 6.4 reduces the number of possible spatial distributions to

$$C(3,2) \times C(2,1) + C(3,2) + C(3,3) \times C(3,2) = 12$$

individual randomization problems.

A Normality Example for the Buffalo Crime Data

Again consider the Buffalo crime data. Using the MINITAB NSCORES command, the modified Wilk-Shapiro statistic is 0.980, implying an underlying normal distribution. Since the arson data could be assumed to be sampled from a normally distributed parent population variable, the normality assumption can be invoked, too. Returning to the original MINITAB subroutine constructed earlier, now the attached set of commands becomes:

```
LET C100(K10) = K4
LET K10 = K10 + 1
RANDOM 14 C1;
NORMAL K20 K21.
END
```

Next a macro-subroutine can be created that has the commands

```
COPY C1 C99
LET K20 = MEAN(C1)
LET K21 = STDEV(C1)
RANDOM 14 C1;
NORMAL K20 K21.
LET K10 = 1
EXEC 'SAMPLING' 100
END
```

Again the commands HISTOGRAM and DESCRIBE will summarize the resulting sample sampling distribution. For the random samples drawn here, the resulting distribution is as follows:

interval midpoint	**frequency**	**interval midpoint**	**frequency**
−.4	1	.0	24
−.3	9	.1	9
−.2	27	.2	3
−.1	24	.3	2
$\hat{\mu}_{MC} = -0.0841$	$(\mu_{MC} = -0.0769)$	.4	1
$\hat{\sigma}_{MC} = 0.1444$			

Using the NSCORES command, the 100 sampled values appear to conform very closely to a normal distribution (the modified Wilk-Shapiro statistic is 0.985).

Standard Errors and Significance Tests

Returning to Figure 5.1, and the case of nominal data, the sampling distributions for WW, BW, and BB are constructed by calculating these counts for each possible surface, and then constructing a frequency distribution of each type of count. Given the randomization perspective, there are 4!/(2!2!), or six possible spatial distributions when $n_1 = n_2 = 2$. Consequently for this illustration

$$\sigma_{BB}^2 = [2(0 - 2/3)^2 + 4(1 - 2/3)^2]/6 = 2/9\,,$$
$$\sigma_{WW}^2 = 2/9\,, \text{ and}$$
$$\sigma_{BW}^2 = [4(2 - 8/3)^2 + 2(4 - 8/3)^2]/6 = 8/9\,.$$

General expressions for these variances appear in Silk (1979), but apply only to surface partitionings where $n \geq 8$.

Next, given the random sampling assumption, there are 2^4, or sixteen possible spatial distributions, when $n_1 = n_2 = 2$. Consequently, for this illustration

$$\sigma_{WW}^2 = [7(0 - 1)^2 + 4(1 - 1)^2 + 4(2 - 1)^2 + 1(4 - 1)^2]/16 = 5/4\,,$$
$$\sigma_{BB}^2 = 5/4\,, \text{ and}$$
$$\sigma_{BW}^2 = [2(0 - 2)^2 + 12(2 - 2)^2 + 2(4 - 2)^2]/16 = 1\,.$$

Formulae for these variances may be derived analytically from the underlying binomial theory, and in general are as follows:

$$\sigma_{BB}^2 = \mu_{BB} + \sum_{i=1}^{i=n} [\sum_{j=1}^{j=n} c_{ij} (\sum_{j=1}^{j=n} c_{ij} - 1)](n_1/n)^3 -$$
$$\{\sum_{i=1}^{i=n}\sum_{j=1}^{j=n} c_{ij}/2 + \sum_{i=1}^{i=n} [\sum_{j=1}^{j=n} c_{ij} (\sum_{j=1}^{j=n} c_{ij} - 1)]\}(n_1/n)^4\,,$$

$$\sigma_{WW}^2 = \mu_{WW} + \sum_{i=1}^{i=n} [\sum_{j=1}^{j=n} c_{ij} (\sum_{j=1}^{j=n} c_{ij} - 1)](n_2/n)^3 -$$
$$\{\sum_{i=1}^{i=n}\sum_{j=1}^{j=n} c_{ij}/2 + \sum_{i=1}^{i=n} [\sum_{j=1}^{j=n} c_{ij} (\sum_{j=1}^{j=n} c_{ij} - 1]\}(n_2/n)^4\,, \text{ and}$$

$$\sigma_{BW}^2 = \{\sum_{i=1}^{i=n}\sum_{j=1}^{j=n} c_{ij} + \sum_{i=1}^{i=n} [\sum_{j=1}^{j=n} c_{ij} (\sum_{j=1}^{j=n} c_{ij} - 1)]\}(n_1/2)(n_2/n) -$$
$$\{2\sum_{i=1}^{i=n}\sum_{j=1}^{j=n} c_{ij} + 4\sum_{i=1}^{i=n} [\sum_{j=1}^{j=n} c_{ij} (\sum_{j=1}^{j=n} c_{ij} - 1)]\}(n_1/n)^2(n_2/n)^2.$$

Making the appropriate substitutions from Figure 5.1 into these formulae yield the three aforementioned numerical values.

Meanwhile, as the foregoing examples suggest, although the expected values of GR and MC do not differ between sampling assumptions, their standard errors do vary. The simplest versions of the standard errors are for the normality assumption. Here:

$$\sigma_{GR}^2 = [\{\sum_{i=1}^{i=n}\sum_{j=1}^{j=n} c_{ij} + \sum_{i=1}^{i=n}[\sum_{j=1}^{j=n} c_{ij}(\sum_{j=1}^{j=n} c_{ij} - 1)]/2\}(n-1) - (\sum_{i=1}^{i=n}\sum_{j=1}^{j=n} c_{ij})^2/2]\,[(n+1)(\sum_{i=1}^{i=n}\sum_{j=1}^{j=n} c_{ij}/2)^2]\text{, and}$$

$$\sigma_{MC}^2 = [2n^2\sum_{i=1}^{i=n}\sum_{j=1}^{j=n} c_{ij} - 4n\{\sum_{i=1}^{i=n}\sum_{j=1}^{j=n} c_{ij} + \sum_{i=1}^{i=n}[\sum_{j=1}^{j=n} c_{ij}(\sum_{j=1}^{j=n} c_{ij} - 1)]\} + 3(\sum_{i=1}^{i=n}\sum_{j=1}^{j=n} c_{ij})^2]/[(\sum_{i=1}^{i=n}\sum_{j=1}^{j=n} c_{ij})^2(n^2 - 1)] - [-1/(n-1)]^2 .$$

Recalling the parameter values of $\sigma_{GR}^2 \simeq .054$ and $\sigma_{MC}^2 \simeq .096$, that were calculated previously, one finds about an 8% difference. This difference is attributable to the deviation of the frequency distribution for the x_i values from a perfectly normal distribution. This deviation was slight, and so the error here is slight.

As n increases the frequency distributions of GR and MC approach a normal distribution. The randomization assumption for GR generated 12 spatial distributions, and the accompanying sampling distribution (Figure 6.3) has a Shapiro-Wilk statistic of $.6080 < W_{12,.01} = .805$ associated with it. Meanwhile, the normality assumption for GR generated 78 spatial distributions, and the accompanying sampling distribution (Figure 6.4) has a Kolmogorov-Smirnov statistic of $0.1127 < K-S_{.10,78} = 0.1212$ associated with it. Similarly, the randomization assumption for MC has a Shapiro-Wilk statistic of .3040 associated with it, whereas the normality assumption also has a Kolmogorov-Smirnov statistic of .1127 associated with it. Hence, the larger sample context of the normality assumption would lead one to fail to reject the null hypothesis of a normal distribution in the population, while the smaller sample context of the randomization assumption would lead one to reject this null hypothesis.

Because the standard error formulae do not use any sample statistics, and since normality is achieved so quickly, a z test can be used to evaluate observed spatial autocorrelation statistics, under the assumption of normality. Hence,

$$z = (GR - \mu_{GR})/\sigma_{GR} = (GR - 1)/\sigma_{GR}\text{, and}$$
$$z = (MC - \mu_{MC})/\sigma_{MC} = [MC + 1/(n-1)]/\sigma_{MC} .$$

Recalling that GR is inversely related to the clustering of similar values on a map, if one desires to affiliate positive z scores with positive spatial autocorrelation, then its expression can be rewritten as

$$z = -(GR - 1)/\sigma_{GR} = (1 - GR)/\sigma_{GR} .$$

Obviously, if the researcher feels that n is too small to justify using z, then a t-statistic could be employed. However, the evidence presented here suggests that such a conservative approach often will not be necessary.

The randomization assumption yields more complex versions of these two standard error formulae. Here,

$$\sigma_{GR}^2 = [(\sum_{i=1}^{i=n}\sum_{j=1}^{j=n} c_{ij})^2 [-(n-1)^2 m_4/s^4 + (n^2 - 3)]/2 +$$
$$(\sum_{i=1}^{i=n}\sum_{j=1}^{j=n} c_{ij})(n-1)[-(n-1)\, m_4/s^4 + n^2 - 3n + 3] +$$
$$\{\sum_{i=1}^{i=n}\sum_{j=1}^{j=n} c_{ij} + \sum_{i=1}^{i=n} [\sum_{j=1}^{j=n} c_{ij} (\sum_{j=1}^{j=n} c_{ij} - 1)]\}(n-1)[(n^2-n+2)\, m_4/s^4$$
$$- (n^2 + 3n - 6)]/2] /[n(n-2)(n-3)(\sum_{i=1}^{i=n}\sum_{j=1}^{j=n} c_{ij})^2/2]\,,$$
$$\text{where } m_4 = \sum_{i=1}^{i=n} (x_i - \bar{x})^4/_n \text{ and } s^4 = [\sum_{i=1}^{i=n} (x_i - \bar{x})^2/_n]^2 .$$

Similarly,

$$\sigma_{MC}^2 = [2n[(n^2-3n+3) \sum_{i=1}^{i=n}\sum_{j=1}^{j=n} c_{ij}] - 4\{\sum_{i=1}^{i=n}\sum_{j=1}^{j=n} c_{ij} +$$
$$\sum_{i=1}^{i=n} [\sum_{j=1}^{j=n} c_{ij} (\sum_{j=1}^{j=n} c_{ij} - 1)]\}n^2 + 3n(\sum_{i=1}^{i=n}\sum_{j=1}^{j=n} c_{ij})^2$$
$$- (m_4/s^4)\, [2(n^2-n) \sum_{i=1}^{i=n}\sum_{j=1}^{j=n} c_{ij} - 8n\{\sum_{i=1}^{i=n}\sum_{j=1}^{j=n} c_{ij}$$
$$+ \sum_{i=1}^{i=n} [\sum_{j=1}^{j=n} c_{ij} (\sum_{j=1}^{j=n} c_{ij} - 1)]\}$$
$$+ 6(\sum_{i=1}^{i=n}\sum_{j=1}^{j=n} c_{ij})^2]] /[(n-1)(n-2)(n-3)(\sum_{i=1}^{i=n}\sum_{j=1}^{j=n} c_{ij})^2]$$
$$- [-1/(n-1)]^2 .$$

Now, because m_4 and s^4 are constant over all permutations of the x_i values, no constraints due to parameter estimation are placed upon the calculations of σ_{GR}^2 and σ_{MC}^2. Therefore the appropriate test statistic will be z. Again, if the researcher feels that n is too small to justify using z, then a t-statistic could be employed.

Returning to Figure 5.4, where $n = 9$, the null hypothesis one wants to test asks whether or not there is non-zero spatial autocorrelation present in the population. In other words, is the sampling distribution under either randomization or the normality assumption the sampling distribution from which the observed spatial distribution was selected? If test statistics indicate that the answer to this question is 'no,' then one wants to infer that $\mu_{MC} \neq -1/(n-1)$ and $\mu_{GR} \neq 1$, rather than saying that the observed MC and GR values lie in the tails of the sampling distribution, and hence are unrepresentative of a population in which $\mu_{MC} = -1/(n-1)$ and $\mu_{GR} = 1$. The pair of null and alternate hypotheses that generally would be tested are

$$H_O: \mu_{MC} = -1/(n-1) \qquad H_O: \mu_{GR} = 1$$
$$H_A: \mu_{MC} \neq -1/(n-1) \qquad H_A: \mu_{GR} \neq 1$$

The test statistics are under randomization are, for the MC values associated with Figure 5.4,

$$z_{MC,5.4a} = [1/2 - (-1/8)]/\sqrt{7/128} = 2.6726\,,$$
$$z_{MC,5.4b} = [-1/4 - (-1/8)/\sqrt{7/128} = -.5345, \text{ and}$$
$$z_{MC,5.4c} = [-7/8 - (-1/8)]/\sqrt{7/128} = -3.207.$$

Meanwhile, the corresponding test statistics under the normality assumption require that an inference be drawn regarding the normality of the data. The assumption of identically distributed x_i values means that any observed set $\{x_i\}$ can be treated as a single sample of size n from the same distribution, since the n individual distributions cannot be distinguished between. Accordingly, for the three examples of Figure 5.4,

$$b = .5888\,|5-1| + .3244\,|4-2| + .1976\,|4-2| + .0947\,|3-3| + 0.0\,|3-3|$$
$$= 3.3992\,, \text{ and thus}$$

$$W = (3.3992)^2/\sum_{i=1}^{i=9} (x_i - 3)^2 = .9629\,.$$

This value of $W > W_{.10,9} = .859$ implies one would fail to reject the null hypothesis that the x_i values were drawn from a normally distributed population. Unfortunately, since $W \neq 1$, the formulae for $\sigma_{MC}{}^2$ and $\sigma_{GG}{}^2$ will be slightly polluted by sampling error. Nevertheless,

$$z_{MC,5.4a} = [1/2 - (-1/8)]/\sqrt{17/320} = 2.7116,$$
$$z_{MC,5.4b} = [-1/4 - (-1/8)]/\sqrt{17/320} = -.5423, \text{ and}$$
$$z_{MC,5.4c} = [-7/8 - (-1/8)]/\sqrt{17/320} = -3.2540.$$

Parallel test statistics for the GR values under randomization are

$$z_{GR,5.4a} = -[1/3 - 1]/\sqrt{251/4536} = 2.8341,$$
$$z_{GR,5.4b} = -[1 - 1]/\sqrt{251/4536} = 0, \text{ and}$$
$$z_{GR,5.4c} = -[11/6 - 1]/\sqrt{251/4536} = -3.5426.$$

Corresponding calculations under the normality assumption are

$$z_{GR,5.4a} = -\ [1/3 - 1]/\sqrt{1/18} = 2.8284,$$
$$z_{GR,5.4b} = -\ [1 - 1]/\sqrt{1/18} = 0, \text{ and}$$
$$z_{GR,5.4c} = -\ [11/6 - 1]/\sqrt{1/18} = -3.5355.$$

If the z statistic is selected as the test criterion, then for a two-tailed situation $z_{.10} = \pm 1.645$, $z_{.05} = \pm 1.96$, and $z_{.01} = \pm 2.575$. In the case of Figures 5.4a and 5.4c, on the basis of both MC and GR values, considering either the randomization or normality assumption, one would reject the null hypothesis of zero spatial autocorrelation in the population. This statistical decision implies that the observed spatial distribution furnishes evidence to indicate that the spatial distribution of x_i values in the population does not have zero spatial autocorrelation. Depending upon the significance level used, this inference rests upon the probability that only 10%, 5%, or 1% of the time one would expect a z statistic of the observed magnitude, strictly due to sampling error, when in fact the spatial distribution of values in the population exhibits zero spatial autocorrelation. These results are exactly what were expected because of the contrived nature Figure 5.4.

In the meantime, for the case of Figure 5.4b, on the basis of both MC and

GR values, again considering either the randomization or the normality assumption, one would fail to reject the null hypothesis. Moreover, the observed spatial distribution supplies no evidence to indicate that the spatial distribution of x_i values in the population is other than random. Now, depending upon the significance level used, 90%, 95% or 99% of the time one would expect a z statistic of the observed magnitude, strictly due to sampling error, when in fact the spatial distribution of values in the population exhibits zero spatial autocorrelation. In other words, any deviation from randomness can be attributed solely to sampling error. Once more this result is not surprisng, because of the contrived nature of Figure 5.4.

In terms of the Buffalo crime data, the accompanying z statistic under the randomization assumption is

$$z = [-0.25497 - (-1/13)]/0.1503 = -1.1844 \quad .$$

Using the normality assumption, this statistic is:

$$z = [-0.25497 - (-1/13)]/.1475 = -1.2070 \quad .$$

Even at a 10% level of significance, the observed MC does not suggest the presence of spatial autocorrelation.

7

Rewriting Classical Statistical Models for Geographic Analyses

Until this chapter matrix notation has been completely avoided in the algebraic and statistical developments that have been presented. Unfortunately, to do justice to the statistical models that are being reformulated here, some matrix notation must be employed.

The preceding chapters have focused on univariate spatial data series. One natural next step is to ask questions about multivariate spatial data series. The simplest questions address the effects of autocorrelation on correlation, simple bivariate regression, and analysis of variance. These are topics commonly covered near the end of an introductory statistical methods in geography course. Accordingly, attention in this chapter will be restricted to them. In keeping with the foregoing discussion, attention will also be restricted to the simultaneous model of spatial dependency.

Bivariate Regression: The Fixed-effects Model

The first model to receive extensive treatment was the regression model (Cliff and Ord 1973). Let variables $X_O \equiv 1$ and X_1 take on only fixed values. Consider the linear regression model

$$\mathbf{Y} = \mathbf{X}\boldsymbol{\beta} + \boldsymbol{\xi} \tag{7.1}$$

where $\mathbf{Y}$ = an n-by-1 vector of values to be predicted (*i.e.*, the dependent variable),

$\mathbf{X}$ = an n-by-2 vector of values from which a prediction is to be made [*i.e.*, the independent variable coupled with $X_O \equiv 1$ in order to estimate an intercept],

$\boldsymbol{\beta}$ = an 2-by-1 vector of regression parameters, and

$\boldsymbol{\xi}$ = an n-by-1 vector of random error terms.

With regard to spatial autocorrelation (Fisher 1971; Griffith 1976), all terms in equation (7.1) may be pre-multiplied by the spatial autocorrelation parameter, ρ_*, and the stochastic connectivity matrix **W**, producing

$$\rho_*\mathbf{WY} = \rho_*\mathbf{WX}\boldsymbol{\beta} + \rho_*\mathbf{W}\boldsymbol{\xi} \tag{7.2}$$

Subtracting equation (7.2) from (7.1) yields

$$\mathbf{Y} - \rho_* \mathbf{WY} = (\mathbf{X}\boldsymbol{\beta} - \rho_* \mathbf{WX}\boldsymbol{\beta}) + (\boldsymbol{\xi} - \rho_* \mathbf{W}\boldsymbol{\xi}) \text{ or}$$
$$(\mathbf{I} - \rho_* \mathbf{W})\mathbf{Y} = (\mathbf{I} - \rho_* \mathbf{W})\mathbf{X}\boldsymbol{\beta} + (\mathbf{I} - \rho_* \mathbf{W})\boldsymbol{\xi} \quad . \tag{7.3}$$

After Griffith (1986), two distinct cases of equation (7.3) will be considered here, namely the variable ξ is autocorrelated, and the variable Y is autocorrelated. Only the case of autocorrelated errors will be discussed in this section. An autocorrelated Y variate will be taken up, in part, in the next section.

In the first case, let $\rho_* = \rho_\xi$, yielding

$$\boldsymbol{\xi} = \rho_\xi \mathbf{W}\boldsymbol{\xi} + \boldsymbol{\Omega} \quad \text{, or}$$
$$\boldsymbol{\Omega} = (\mathbf{I} - \rho_\xi \mathbf{W})\boldsymbol{\xi} \quad , \tag{7.4}$$

where Ω is an independent, random normal variate of constant variance.

Substituting equation (7.4) into (7.3) gives

$$(\mathbf{I} - \rho_\xi \mathbf{W})\mathbf{Y} = (\mathbf{I} - \rho_\xi \mathbf{W})\mathbf{X}\boldsymbol{\beta} + (\mathbf{I} - \rho_\xi \mathbf{W})\boldsymbol{\xi} \quad \text{, or}$$
$$\mathbf{Y} = \mathbf{X}\boldsymbol{\beta} + (\mathbf{I} - \rho_\xi \mathbf{W})^{-1} \boldsymbol{\Omega} \quad . \tag{7.5}$$

Parameter estimates for equation (7.5) are

$$\hat{\boldsymbol{\beta}} = [\mathbf{X}^T(\mathbf{I} - \rho_\xi \mathbf{W})^T(\mathbf{I} - \rho_\xi \mathbf{W})\mathbf{X}]^{-1}\mathbf{X}^T(\mathbf{I} - \rho_\xi \mathbf{W})^T(\mathbf{I} - \rho_\xi \mathbf{W})\mathbf{Y} \quad \text{, and}$$
$$\hat{\sigma}^2 = (\mathbf{Y} - \mathbf{X}\boldsymbol{\beta})^T(\mathbf{I} - \rho_\xi \mathbf{W})^T(\mathbf{I} - \rho_\xi \mathbf{W})(\mathbf{Y} - \mathbf{X}\boldsymbol{\beta})/(n - 2) \quad ,$$

where the superscript T refers to matrix transpose.

The standard errors of these regression parameter estimates are given by

$$[\mathbf{X}^T(\mathbf{I} - \rho_\xi \mathbf{W})^T(\mathbf{I} - \rho_\xi \mathbf{W})\mathbf{X}]^{-1}\sigma^2 \quad .$$

A simple closed form solution does not exist for the standard errors of $\hat{\sigma}^2$ and $\hat{\rho}_\xi$.

For classical inferential purposes the estimation of $\boldsymbol{\beta}$, σ^2 and ρ_ξ requires an assumption of normally distributed error terms, and a nonlinear iterative procedure for estimating ρ_ξ and the other two parameters. If $p_\xi = 0$, the expected value term reduces to $E(\boldsymbol{\Omega}^T\boldsymbol{\Omega}) = \sigma^2_\Omega \mathbf{I}$, and hence $\text{var}(\hat{\boldsymbol{\beta}}) = \sigma^2_\Omega(\mathbf{X}^T\mathbf{X})^{-1}$. On the other hand, though, if $\rho^2_\xi \neq 0$, then this expectation does not reduce. Consequently, if the error term ξ of equation (7.1) is spatially autocorrelated, then the conventional estimate of the standard error of $\boldsymbol{\beta}$ is biased. Accordingly, resulting t-tests could easily lead to incorrect statistical decisions for the null hypothesis H_o: $\boldsymbol{\beta} = \mathbf{0}$.

The Analysis of Variance Model

Suppose $\rho_* = \rho_Y$, yielding

$$\mathbf{Y} = \rho_Y \mathbf{WY} + \mathbf{Y}^* \quad \text{, or}$$
$$\mathbf{Y}^* = (\mathbf{I} - \rho_Y \mathbf{W})\mathbf{Y} \quad , \tag{7.6}$$

where Y^* is an independent, random variable that can be predicted from some set $\{X_j; j=0,1, \ldots ,p-1)$.

Substituting equation (7.6) into equation (7.3) yields

$$\mathbf{Y}^* = (\mathbf{I} - \rho_Y \mathbf{W})\mathbf{X}\boldsymbol{\beta} + (\mathbf{I} - \rho_Y \mathbf{W})\boldsymbol{\xi} \quad . \tag{7.7}$$

Now using this conception of variate Y, the analysis of variance model can be rewritten as a regression problem in which the X variables are fixed-effects dummy variables indicating whether or not an areal unit is in a particular region. The accompanying F-ratio is given by (Griffith 1978):

$$\frac{[\hat{\beta}^T \mathbf{X}^T(\mathbf{I} - \rho_y\mathbf{W})^T(\mathbf{I} - \rho_y\mathbf{W})\mathbf{Y}]/p}{[\mathbf{Y}^T(\mathbf{I} - \rho_y\mathbf{W})^T(\mathbf{I} - \rho_y\mathbf{W})\mathbf{Y} - \hat{\beta}^T\mathbf{X}^T(\mathbf{I} - \rho_y\mathbf{W})^T(\mathbf{I} - \rho_y\mathbf{W})\mathbf{Y}]/(n - p)} \quad . \tag{7.8}$$

If $\rho > 0$, then the spatial linear operator $(\mathbf{I} - \rho_y\mathbf{W})$ means that each areal unit value is being reduced by a weighted average of its neighboring values. Hence the variance terms will tend to have the following relation:

$$\mathbf{Y}^T(\mathbf{I} - \rho_y\mathbf{W})^T(\mathbf{I} - \rho_y\mathbf{W})\mathbf{Y} < \mathbf{Y}^T\mathbf{Y} \quad .$$

Intuitively this relationship makes sense because positive spatial autocorrelation dampens local variation. Combining this result with expression (7.8) implies that as $\rho \rightarrow 1$, the regional differences between groups of y_i values tend to increase, even though the underlying autocorrelated errors do not display significant regional differences. Moreover, because the total variation is being reduced, smaller deviations in the regional variances become more pronounced, relatively speaking. Thus, the probability of rejecting the null hypothesis when it is true tends to increase. On the other hand, as $\rho \rightarrow -1$, the regional differences between groups of y_i values tend to decrease, even though the underlying autocorrelated errors do display significant regional differences. Local map variation is being accentuated, for the spatial linear operator has become $(\mathbf{I} + |\rho_y|\mathbf{W})$. Accordingly, each areal unit value is being incremented by a weighted average of its neighbors. Now the tendency for the relation between variance terms is

$$\mathbf{Y}^T\mathbf{Y} < \mathbf{Y}^T(\mathbf{I} - \rho_y\mathbf{W})^T(\mathbf{I} - \rho_y\mathbf{W})\mathbf{Y} \quad .$$

Moreover, because the total variation is being increased, larger deviations in the regional variances become less pronounced, relatively speaking. Hence the probability of failing to reject the null hypothesis when it is false tends to increase.

Finally, evidence exists to indicate that variance heterogeneity and deviations from normality, at least in part, can be attributable to the presence of non-zero spatial autocorrelation (Griffith 1981a). More will be said about these two situations in the next chapter.

Bivariate Regression: The Variable-Effects Model

A third autocorrelation case arises when the X values are random variables that also may be spatially autocorrelated. Assuming bivariate normality, the parameters to estimate now are ρ_x, ρ_y, σ_x, σ_y, μ_x, μ_y and ρ. If $\rho = 0$ then $\beta = 0$, and this case reduces to that of two independent univariate autocorrelation problems. If $\rho_x = \rho_y$ then this case reduces to one that exactly parallels the bivariate fixed-effects model. Otherwise the nonlinear estimation problem is far more complicated here, since both ρ_x and ρ_y must be estimated, and since σ_x and σ_y, and μ_x are not independent. The complexities involved here may be exemplified by considering the parameter estimate

$$\hat{\mu}_x = \mathbf{1}^T(\mathbf{I}-\rho_x\mathbf{W})^T[\sigma_y(\mathbf{I}-\rho_x\mathbf{W})\mathbf{X}-\rho\sigma_x(\mathbf{I}-\rho_y\mathbf{W})(\mathbf{Y}-\mu_y\mathbf{1})]/\mathbf{1}^T(\mathbf{I}-\rho_x\mathbf{W})^T(\mathbf{I}-\rho_x\mathbf{W})\mathbf{1}\sigma_y .$$

Now, if $\rho = 0$,

$$\hat{\mu}_x = \mathbf{1}^T(\mathbf{I}-\rho_x\mathbf{W})^T(\mathbf{I}-\rho_x\mathbf{W})\mathbf{X}/\mathbf{I}^T(\mathbf{I}-\rho_x\mathbf{W})^T(\mathbf{I}-\rho_x\mathbf{W})\mathbf{1} \quad ,$$

as was seen earlier and mentioned above. If $\rho_x = \rho_y$, then

$$\hat{\mu}_x = \mathbf{1}^T(\mathbf{I}-\rho_x\mathbf{W})^T(\mathbf{I}-\rho_x\mathbf{W})[\mathbf{X}-\rho\sigma_x(\mathbf{Y}-\mu_y\mathbf{1})/\sigma_y]/\mathbf{1}^T(\mathbf{I}-\rho_x\mathbf{W})^T(\mathbf{I}-\rho_x\mathbf{W})\mathbf{1} \quad ,$$

where $\rho\sigma_x/\sigma_y$ is a standardized regression coefficient, as was mentioned above, Finally, $\hat{\sigma}_x$ is the positive root of

$$\sigma_x^2 + \sigma_x\{\rho(\mathbf{X}-\mu_x\mathbf{1})^T(\mathbf{I}-\rho_x\mathbf{W})^T(\mathbf{I}-\rho_y\mathbf{W})(\mathbf{Y}-\mu_y\mathbf{1})/[n\sigma_y(1-\rho^2)]\} - (\mathbf{X}-\mu_x\mathbf{1})^T(\mathbf{I}-\rho_x\mathbf{W})^T(\mathbf{I}-\rho_x\mathbf{W})(\mathbf{X}-\mu_x\mathbf{1})/[n(1-\rho^2)] = 0 \quad .$$

More general multivariate spatial autocorrelation discussions may be found in Streitberg (1979) and Wartenberg (1985).

Product Moment Correlation

The fourth classical statistical model to be rewritten here is closely affiliated with the preceding bivariate regression model, appearing in it as parameter ρ. Griffith (1980) argues that the presence of non-zero spatial autocorrelation affects classical measures of association between variables. Bivand (1984) corroborates this contention through a set of simulation experiments.

In the presence of spatial autocorrelation, the rewritten estimate of the product moment correlation coefficient ρ becomes

$$[(\mathbf{X}-\mu_x\mathbf{1})^T(\mathbf{I}-\rho_x\mathbf{W})^T(\mathbf{I}-\rho_y\mathbf{W})(\mathbf{Y}-\mu_y\mathbf{1})]/\{[(\mathbf{X}-\mu_x\mathbf{1})^T(\mathbf{I}-\rho_x\mathbf{W})^T(\mathbf{I}-\rho_x\mathbf{W})(\mathbf{X}-\mu_x\mathbf{1})] \times [(\mathbf{Y}-\mu_y\mathbf{1})^T(\mathbf{I}-\rho_y\mathbf{W})^T(\mathbf{I}-\rho_y\mathbf{W})(\mathbf{Y}-\mu_y\mathbf{1})]\}^{1/2}$$

Some Simulation Illustrations

MINITAB can be used to illustrate in a classroom setting the revisions outlined in this chapter, especially using its PC compatible version. Consider a 5-by-5 regular lattic tessellation. The map and connectivity matrix, say M1, for this set of areal units is given in Figure 7.1. Five normal random variables (K1 = 31, 32, . . . ,35) can be generated with the set of commands

```
RANDOM 25 C31 – C35;
NORMAL 0 1.
```

Using the command

```
COPY M1 C1-C25
```

followed by the command:

```
RSUM C1-C25 C26
```

and then the set of commands:

A	B	C	D	E
F	G	H	I	J
K	L	M	N	O
P	Q	R	S	T
U	V	W	X	Y

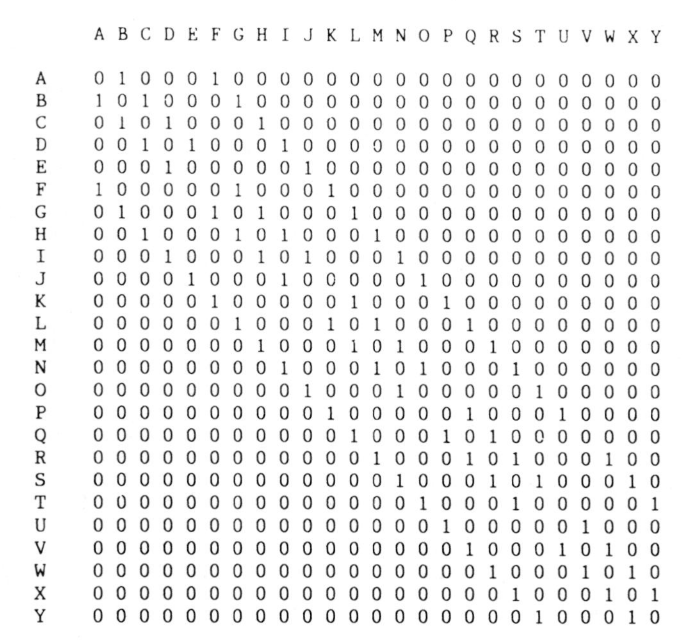

	A	B	C	D	E	F	G	H	I	J	K	L	M	N	O	P	Q	R	S	T	U	V	W	X	Y
A	0	1	0	0	0	1	0	0	0	0	0	0	0	0	0	0	0	0	0	0	0	0	0	0	0
B	1	0	1	0	0	0	1	0	0	0	0	0	0	0	0	0	0	0	0	0	0	0	0	0	0
C	0	1	0	1	0	0	0	1	0	0	0	0	0	0	0	0	0	0	0	0	0	0	0	0	0
D	0	0	1	0	1	0	0	0	1	0	0	0	0	0	0	0	0	0	0	0	0	0	0	0	0
E	0	0	0	1	0	0	0	0	0	1	0	0	0	0	0	0	0	0	0	0	0	0	0	0	0
F	1	0	0	0	0	0	1	0	0	0	1	0	0	0	0	0	0	0	0	0	0	0	0	0	0
G	0	1	0	0	0	1	0	1	0	0	0	1	0	0	0	0	0	0	0	0	0	0	0	0	0
H	0	0	1	0	0	0	1	0	1	0	0	0	1	0	0	0	0	0	0	0	0	0	0	0	0
I	0	0	0	1	0	0	0	1	0	1	0	0	0	1	0	0	0	0	0	0	0	0	0	0	0
J	0	0	0	0	1	0	0	0	1	0	0	0	0	0	1	0	0	0	0	0	0	0	0	0	0
K	0	0	0	0	0	1	0	0	0	0	0	1	0	0	0	1	0	0	0	0	0	0	0	0	0
L	0	0	0	0	0	0	1	0	0	0	1	0	1	0	0	0	1	0	0	0	0	0	0	0	0
M	0	0	0	0	0	0	0	1	0	0	0	1	0	1	0	0	0	1	0	0	0	0	0	0	0
N	0	0	0	0	0	0	0	0	1	0	0	0	1	0	1	0	0	0	1	0	0	0	0	0	0
O	0	0	0	0	0	0	0	0	0	1	0	0	0	1	0	0	0	0	0	1	0	0	0	0	0
P	0	0	0	0	0	0	0	0	0	0	1	0	0	0	0	0	1	0	0	0	1	0	0	0	0
Q	0	0	0	0	0	0	0	0	0	0	0	1	0	0	0	1	0	1	0	0	0	0	0	0	0
R	0	0	0	0	0	0	0	0	0	0	0	0	1	0	0	0	1	0	1	0	0	0	1	0	0
S	0	0	0	0	0	0	0	0	0	0	0	0	0	1	0	0	0	1	0	1	0	0	0	1	0
T	0	0	0	0	0	0	0	0	0	0	0	0	0	0	1	0	0	0	1	0	0	0	0	0	1
U	0	0	0	0	0	0	0	0	0	0	0	0	0	0	0	1	0	0	0	0	0	1	0	0	0
V	0	0	0	0	0	0	0	0	0	0	0	0	0	0	0	0	1	0	0	0	1	0	1	0	0
W	0	0	0	0	0	0	0	0	0	0	0	0	0	0	0	0	0	1	0	0	0	1	0	1	0
X	0	0	0	0	0	0	0	0	0	0	0	0	0	0	0	0	0	0	1	0	0	0	1	0	1
Y	0	0	0	0	0	0	0	0	0	0	0	0	0	0	0	0	0	0	0	1	0	0	0	1	0

FIGURE 7.1 The Planar Partitioning and Connectivity Matrix for a 5-by-5 Regular Lattice

```
LET K1 = 1
STORE
LET CK1 = CK1/C26
LET K1 = K1 + 1
END
EXECUTE 25 TIMES
COPY C1-C25 M1
```

will convert matrix **C** into matrix **W.** The command:

```
CORR C31-C35
```

will calculate the product moment correlation coefficient between the independent variables. The commands:

```
SET C30
25(1)
END
LET K1 = 32
LET K2 = 31
STORE 'REG'
REGRESS CK2 2 C30 CK1;
NOCONSTANT.
LET K1 = K1 + 1
END
EXECUTE 'REG' 4 TIMES
```

will yield four sets of bivariate regression estimates. Finally, if the geographic landscape is divided into four regions, say {A,B,C,F,G,H,K,L,M}, {D,E,I,J,N,O}, {P,Q,R,U,V,W} and {S,T,X,Y}, the corresponding analysis of variance can be obtained by establishing the following classification variable:

```
SET C40
3(1) 2(2) 3(1) 2(2) 3(1) 2(2) 3(3) 2(4) 3(3) 2(4)
END
```

The analysis is executed with the commands:

```
LET K1 = 31
STORE 'ANOVA'
ONEWAY CK1 C40
LET K1 = K1 + 1
END
EXECUTE 'ANOVA' 5 TIMES
```

These computer codes yield results for the classical statistics situation.

Next, the variables C31 through C35 need to have spatial autocorrelation embedded into them. Let C31 be Y, as in the regression analysis above. Consider the values of $\rho_y = 0.9$, and $\rho_x = 0.0, 0.3, 0.6$ and 0.9. Then the autocorrelated variables can be constructed using the following commands:

```
DIAGONAL C30 M2
LET K1 = 0.9
LET C42 = C32
STORE
MULTIPLY M1 K1 M3
SUBTRACT M3 FROM M2 M4
END
EXECUTE
MULTIPLY M4 C31 C41
MULTIPLY M4 C35 C45
LET K1 = 0.3
EXECUTE
MULTIPLY M4 C33 C43
LET K1 = 0.6
EXECUTE
MULTIPLY M4 C34 C44
```

The comparative results are obtained, then, by repeating the selected foregoing commands, or

```
CORR C41-C45
LET K1 = 42
LET K2 = 41
EXECUTE 'REG' 4 TIMES
LET K1 = 41
EXECUTE 'ANOVA' 5 TIMES
DESCRIBE C31-C35,C41-C45
```

Results of a sample execution of this computer code are as follows (results between executions of the program will more than likely differ because of the MINITAB random number generator routine):

		unautocorrelated			autocorrelated		
variable	ρ	$\hat{\mu}$	$\hat{\sigma}$	F	$\hat{\mu}$	$\hat{\sigma}$	F
C31	0.9	−.17	1.16	1.26	.01	1.37	.28
C32	0.0	−.26	1.07	.31	−.26	1.07	.31
C33	0.3	.04	1.03	4.40	.03	.99	2.25
C34	0.6	.25	.90	1.24	.09	1.04	.39
C35	0.9	−.08	.94	.99	−.03	.94	.19

	unautocorrelated			autocorrelated		
regression model	$\hat{\alpha}$	$\hat{\beta}$	$\hat{\sigma}^2$	$\hat{\alpha}$	$\hat{\beta}$	$\hat{\sigma}^2$
C31 on C32	−.089	.321	1.133	.076	.270	1.367
C31 on C33	−.173	.001	1.187	−.003	.320	1.365
C31 on C34	−.255	.335	1.146	−.030	.387	1.337
C31 on C35	−.166	.089	1.184	.010	.130	1.393

correlation matrix

	C31	C32	C33	C34
C32	.297			
C33	.001	.164		
C34	.259	.002	−.057	
C35	.072	.128	.320	.052

	C41	C42	C43	C44
C42	.212			
C43	.218	.169		
C44	.294	−.027	−.156	
C45	.089	.148	.458	−.253

As these results indicate, noticeable differences exist between the results from unautocorrelated and autocorrelated data.

Estimation When Non-Normality Prevails

Another method of estimating ρ, which differs from the nonlinear optimization one repeatedly mentioned above, is to substitute equation (4.2) into the GR formula, and then set this formula equal to its expected value. In other words, what value would ρ need to be in order for E(GR) = 1 for the ξ_i values, but not for the x_i values? This substitution renders:

$$\frac{(n-1)\sum_{i=1}^{i=n}\sum_{j=1}^{j=n} c_{ij}\left[\left(x_i-\rho\sum_{k=1}^{k=n} w_{ik}x_k\right)-\left(x_j-\rho\sum_{k=1}^{k=n} w_{jk}x_k\right)\right]^2}{\left(2\sum_{i=1}^{i=n}\sum_{j=1}^{j=n} c_{ij}\right)\sum_{i=1}^{i=n}\left[\left(x_i-\rho\sum_{j=1}^{j=n} w_{ij}x_j\right)-\sum_{i=1}^{i=n}\left(x_i-\rho\sum_{j=1}^{j=n} w_{ij}x_j\right)/n\right]^2}=1.$$

This equation may be rewritten as a quadratic equation in ρ (see Appendix F). Its real root that is closest to zero in absolute value is $\hat{\rho}$. If no real root exists, then equating its partial derivative with respect to ρ with zero, and then solving for ρ, will yield $\hat{\rho}$. Of course, the MC index could be used for this same purpose, too (Appendix F).

For the purpose of comparing these estimates, once again consider the spatial distribution I appearing in Figure 6.2. The connectivity matrix **C** for this example is

	A	B	C	D
A	0	1	1	0
B	1	0	0	1
C	1	0	0	1
D	0	1	1	0

The four eigenvalues of matrix **C** are 2, 0, 0, and -2. Hence, the equation (B2) in Appendix B becomes

$$[-1/(1-\rho) + 0 + 0 + 1/(1+\rho)] - (1+\rho)/[(2 + 2\rho + \rho^2)/4] = 0\,,$$

which is a cubic equation having the roots 2.21432, -1.67513 and -0.53919. Substitutions into equation (F1) of Appendix F yield:

$$\rho^2[(3/16)(8) - 1] - 2\,\rho[(3/16)(-8) - (-1)] + [(3/16)(12) - 2] = 0,$$

which is a quadratic equation having the roots -1.70711 and -0.29289. Finally, substitutions into equation (F2) of Appendix F yield:

$$\rho^2\{1 + [(4)(3)/8](-2)\} - \rho\{(2)(-1) + [(4)(3)/8](4)\} + \{2 + [(4)(3)/8](-2)\} = 0,$$

which is a quadratic equation having the roots -1.70711 and -0.29289. Consequently, the appropriate estimates of ρ are

equation (B2): $\rho = -0.53919$,
equation (F1): $\rho = -0.29289$, and
equation (F2): $\rho = -0.29289$.

This simple example illustrates the fact that these different estimating equations may not render the same statistic.

The difference between using the foregoing likelihood function and using the GR or MC substitution is that this former expression requires the assumption of normality, whereas this latter expression can be used in other situations. As the example presented here demonstrates, unfortunately, these estimation procedures fail to yield equivalent results. But the GR and MC substitutions permit one to remain in the realm of linear models, with results for estimates such as μ and σ^2 becoming conditional on the estimate of ρ_*.

8

Implications for Statistical Work in Geography

Three empirical examples will be discussed in this section. The purpose of presenting these examples is to illustrate the utility and importance of spatial autocorrelation analysis to empirical problems. These illustrations aim to highlight issues covered in the preceding chapters.

The 1943/44 Manitoba Grain Handling System

The first example explores the collection of grain in Manitoba during the 1943/44 crop year (Griffith 1982). Data are for 388 grain elevator locations. Because empirical evidence suggests that farmers at that time tended to minimize the hauling distance between their farms and the grain elevator collection points to which they delivered their crops, these points were used to generate a set of Thiessen polygons for the province of Manitoba. The resulting surface partitioning appears in Figure 8.1. Adjustments have been made for boundary conditions as well as the presence of massive lakes in Manitoba's northeast. A Kolmogorov-Smirnov statistic of $0.1175 < \text{K-S}_{.05,\infty}$ was calculated for the affiliated grain receipt data. Therefore, there is no evidence to indicate that the frequency distribution for these data does not conform to a normal distribution in the population. Hence, the standard error expressions under an assumption of normality can be used to test the significance of calculated GR and MC values. The calculated spatial autocorrelation indices were GR = .8230 and MC = .0806, both of which are significant at the 1% level. Consequently, there is evidence to indicate that grain elevators with large receipts tended to have neighboring grain elevators with lower receipts. This finding is consistent with the notion of spatial competition found in economic geography. It also furnishes some evidence in support of the need to guide grain elevator site closures in order to rationalize the system.

1971 Toronto Population Density

The second example concerns the geographical distribution of population density in Toronto (Griffith 1981c). Its purpose is to exemplify the impact of spatial autocorrelation on regression parameter estimates. The model cali-

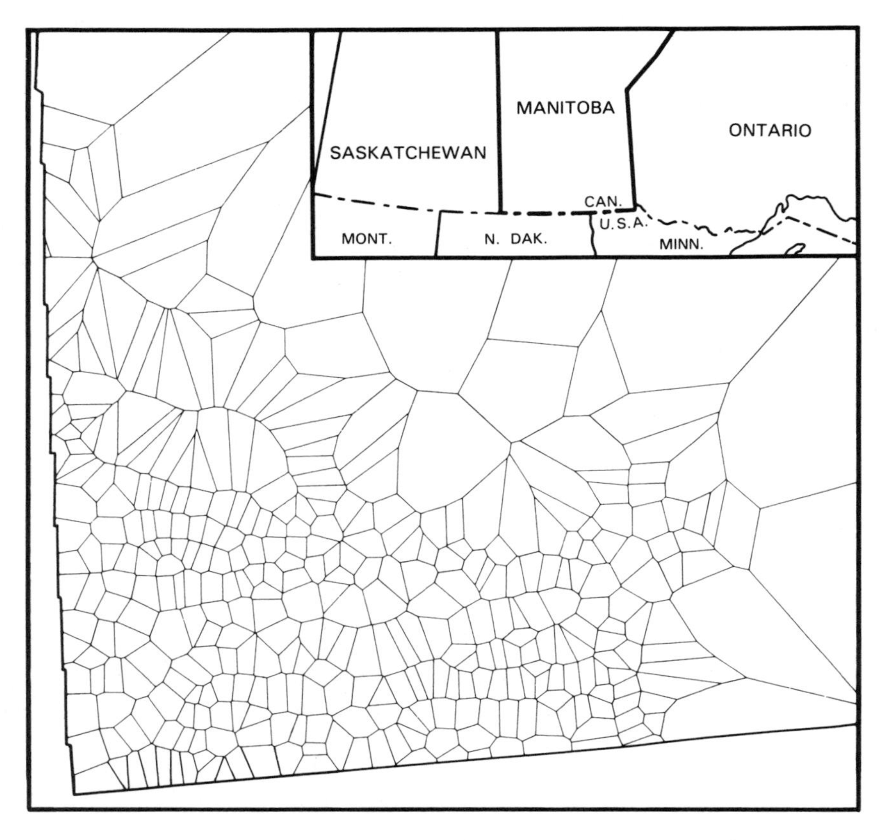

FIGURE 8.1 A Thiessen Polygon Partitioning of the Province of Manitoba Based upon the 1943/44 Geographic Distribution of Grain Elevator Locations (Griffith 1982; reproduced by permission of *Economic Geography)*

brated was of the form

$$D_i = A\,\exp^{(-b\,d_{i,CBD})} \quad , \qquad (8.1)$$

where D_i = the population density of areal unit i,
$d_{i,CBD}$ = the distance separating areal unit i from the CBD, and
A, b = parameters.

A non-linear least squares calibration of equation (8.1) rendered

$$\hat{A} = 26993.73 \quad , \text{and} \quad \hat{b} = .0929 \quad .$$

Rewriting equation (8.1) as the autoregressive model

$$D_i = \rho \sum_{j=1}^{j=346} w_{ij}\,D_j + A\,\exp(-b\,d_{i,CBD})$$

produced the estimates

$$\hat{A} = 10919.66 \quad , \quad \hat{b} = .0989 \quad , \text{and} \quad \hat{\rho} = .7837 \quad .$$

Clearly the parameter estimate $\hat{A}$ is affected dramatically by the presence of non-zero spatial autocorrelation. In contradistinction, though, the distance decay parameter $\hat{b}$ does not seem to be influenced very much by this phenomenon. In other words, ignoring the presence of spatial autocorrelation in the predicted variable can lead to biased parameter estimates.

Agricultural Production in Puerto Rico

The third example deals with the geographical distribution of agricultural production in Puerto Rico. Focusing on a single aspect of this production, a maximum likelihood estimate of ρ for the 1974 spatial distribution of sugarcane harvested was found to be 0.1856 (Griffith 1983). Using the modified Geary Ratio procedure (see Appendix F) to calculate the estimate $\hat{\rho}$ yielded a value of 1.1624 (see Griffith, 1979). Clearly, since $|\rho_{GR}| > 1$, the maximum likelihood estimate of 0.1856 is preferable. This latter finding suggests a weak tendency for similar harvest values to cluster on the map.

An inspection of all Puerto Rican agricultural production in 1969 rendered additional thought–provoking results (Griffith, 1981a). This island is divided into five agricultural administrative regions. The generalized variance in agricultural output across these five regions had an F-ratio of 11.882 for the spatially autocorrelated data, which is significant at the 1% level. Transforming these data to unautocorrelated data using the spatial linear operation $(\mathbf{I} - \rho_{x_j}\mathbf{W})$ resulted in an F-ratio of 1.163 for variance homogeneity, a statistic which is not significantly different from unity. Meanwhile, the test for multivariate normality within each region yielded the following Shapiro-Wilk statistics:

region	**autocorrelated data**	**unautocorrelated data**
San Juan	.9399	.9312
Arecibo	.8312**	.9787
Mayaguez	.9402	.9092
Ponce	.9699	.9681
Caguas	.8495**	.9573

(** denotes a significant difference from 1 at the 5% level)

Consequently, in the cases of the Arecibo and Caguas regions, incorporating the spatial linear operators $(\mathbf{I} - \rho_{x_j}\mathbf{W})$ into the analysis transformed the data $\mathbf{X}$ which deviated significantly from a multivariate normal distribution, to the synthetic data $\mathbf{X}^*$ which did not deviate significantly from normality. Conventional statistical wisdom would have dictated that ill-advised data transformations, such as square root, logarithm, or inverse, be applied to $\mathbf{X}$ in order to try to achieve this end.

Concluding Comments

This monograph has attempted to present the fundamental ideas of spatial autocorrelation in a pedagogic way. Emphasis has been placed on the difference between arrangement properties and random variable properties of a variable, as well as their corresponding sampling distributions under the null hypothesis. One important point focused on in this last section is that many past statistical geographical analyses may be misleading because researchers suggest that parameter estimates are associated with traditional random–variable populations. The principal purpose of this treatise has been to encourage the spatial analyst to remove suspicion about results either by demonstrating that the prevailing level of spatial autocorrelation in the landscape is approximately zero, or by modifying statistical analyses so that spatial autocorrelation is accommodated. To this end, it is hoped that recognizing spatial autocorrelation will circumvent many of the problems that beset quantitative geography.

Bibliography

Textbook Introductions

Ahuja, N. and B. J. Schachter, 1983. *Pattern Models.* New York: Wiley.

Bennett, R. 1979. *Spatial Time Series: Analysis-Forecasting-Control.* New York: Methuen.

Clark, W. and P. Hosking. 1986. *Statistical Methods for Geographers.* New York: Wiley.

Cliff, A. and J. Ord. 1973. *Spatial Autocorrelation.* London: Pion.

Cliff, A. and J. Ord. 1981a. *Spatial Processes: Models and Applications.* London: Pion.

Cliff, A. *et al.* 1975. *Elements of Spatial Structure: A Quantitative Approach.* Cambridge: Cambridge University Press.

Ebdon, D. 1977. *Statistics in Geography: A Practical Approach.* Oxford: Basil Blackwell.

Greer-Wootten, B. 1972. *A Bibliography of Statistical Applications in Geography.* Washington, D.C.: Association of American Geographers, Technical Paper 9.

Griffith, D. 1984. "Theory of Spatial Statistics," pp. 3–15 in G. Gaile and C. Willmott (eds.), *Spatial Statistics and Models.* Boston: D. Reidel.

Haggett, P., A. Cliff and A. Frey. 1977. *Locational Analysis in Human Geography,* vol. 2. London: Edward Arnold.

Johnston, R. 1978. *Multivariate Statistical Analysis in Geography.* New York: Longman.

Mather, P. 1976. *Computational Methods of Multivariate Analysis in Physical Geography.* New York: Wiley.

Paelinck, J. and L. Klaassen. 1979. *Spatial Econometrics.* Farnborough, Eng.: Saxon House.

Silk, J. 1979. *Statistical Concepts in Geography.* London: George Allen Unwin.

Taylor, P. 1977. *Quantitative Methods in Geography: An Introduction to Spatial Analysis.* Boston: Houghton Mifflin Co.

Upton, G. and B. Fingleton. 1985. *Spatial Data Analysis by Example.* New York: Wiley.

Yeates, M. 1974. *An Introduction to Quantitative Analysis in Human Geography.* New York: McGraw-Hill.

Overviews

Arora, S. and M. Brown. 1977. "Alternative Approaches to Spatial Autocorrelation: An Improvement Over Current Practice," *International Regional Science Review* 2, 1:67–78.

Cliff, A. and J. Ord. 1981b. "The Effects of Spatial Autocorrelation on Geographical Modelling," pp. 108–137 in R. Craig and M. Labovitz (eds.), *Future Trends in Geomathematics.* London: Pion.

Cliff, A. and J. Ord. 1981c. "Spatial and Temporal Analysis: Autocorrelation in Space and Time," pp. 104–110 in N. Wrigley and R. Bennett (eds.), *Quantitative Geography: A British View.* London: Routledge and Kegan Paul.

Doreian, P. 1980. "Linear Models With Spatially Distributed Data," *Sociological Methods and Research* 9:29–60.

Gatrell, A. 1979a. "Autocorrelation in Spaces," *Environment and Planning* A 11:507–516.

Griffith, D. 1980. "Towards a Theory of Spatial Statistics," *Geographical Analysis* 12: 325–339.

Griffith, D. 1986. "An Overview of the Spatial Statistics Literature," *Discussion Paper* No. 1986/6. Rotterdam: Netherlands Economic Institute.

Haining, R. 1980. "Spatial Autocorrelation Problems," pp. 1–44 in D. Herbert and R. Johnston (eds.), *Geography and the Urban Environment: Progress in Research and Application,* vol. 3. New York: Wiley.

Haining, R., D. Griffith and R. Bennett. 1983. "Simulating Two-dimensional Autocorrelated Surfaces," *Geographical Analysis* 15:247–255.

Hepple, L. 1974. "The Impact of Stochastic Process Theory Upon Spatial Analysis in Human Geography," *Progress in Geography* 6:89–142.

Hubert, L. *et al.* 1985. "Measuring Association Between Spatially Defined Variables: An Alternative Procedure," *Geographical Analysis* 17:36–46.

Unwin, D. and L. Hepple. 1974. "The Statistical Analysis of Spatial Series," *The Statistician* 23:211–227.

Indices

Bartels, C. and I. Hordijk. 1977. "On the Power of the Generalized Moran Contiguity Coefficient in Testing for Spatial Autocorrelation Among Regression Disturbances," *Regional Science and Urban Economics* 7:83–101.

Cliff, A. and J. Ord. 1975a. "The Choice of a Test for Spatial Autocorrelation," pp. 54–77 in J. Davis and M. McCullagh (eds.), *Display and Analysis of Spatial Data.* New York: Wiley.

Cliff, A., R. Martin and J. Ord. 1975. "A Test for Spatial Autocorrelation in Choropleth Maps Based Upon a Modified χ^2 Statistic," *Transactions and Papers,* Institute of British Geographers 65:109–129.

Dacey, M. 1968. "A Review on Measures of Contiguity for Two- and K-color Maps," pp. 479–495 in B. Berry and D. Marble (eds.), *Spatial Analysis: A Reader in Statistical Geography.* Englewood Cliffs, NJ: Prentice-Hall.

de Jong, P., C. Sprenger and F. van Veen. 1984. "On Extreme Values of Moran's I and Geary's c," *Geographical Analysis* 16:17–24.

Geary, R. 1954. "The Contiguity Ratio and Statistical Mapping," *The Incorporated Statistician* 5:115–145.

Meade, R. 1967. "A Mathematical Model for the Estimation of Interplant Competition," *Biometrics* 23:189–205.

Moran, P. 1948. "The Interpretation of Statistical Maps," *Journal of the Royal Statistical Society* 10B:243–251.

Sen, A. 1976. "Large Sample-size Distribution of Statistics Used in Testing for Spatial Correlation," *Geographical Analysis* 8:175–184.

Sen, A. and S. Soot. 1977. "Rank Tests for Spatial Correlation," *Environment and Planning* A 9:897–903.

Conceptual Treatments

Agterberg, F. 1970. "Autocorrelation Functions in Geology," pp. 113–144 in D. Merriam (ed.), *Geostatistics: A Symposium.* New York: Plenum.

Agterberg, F. 1981. "Cell-value Distribution Models in Spatial Pattern Analysis," pp. 5–28 in R. Craig and M. Labovitz (eds.), *Future Trends in Geomathematics.* London: Pion.

Amrhein, C., J. Guevara and D. Griffith. 1983. "The Effect of Random Thiessen Structure and Random Processes on the Measurement of Spatial Autocorrelation," *Proceedings,* 14th Annual Pittsburgh Modelling and Simulation Conference 14:585–589.

Anselin, L. 1980. *Estimation Methods for Spatial Autoregressive Structures.* Ithaca, NY: Regional Science Dissertation and Monograph Series #8, Cornell University.

Bivand, R. 1984. "Regression Modeling With Spatial Dependence: An Application of Some Class Selection and Estimation Methods," *Geographical Analysis* 16:25–37.

Bodson, P. and D. Peeters. 1975. "Estimation of the Coefficients of a Linear Regression in the Presence of Spatial Autocorrelation: An Application to a Belgian Labour-Demand Function," *Environment and Planning* A 7:455–472.

Brandsma, A. and R. Ketellapper. 1979a. "A Biparametric Approach to Spatial Autocorrelation," *Environment and Planning* A 11:51–58.

Brandsma, A. and R. Ketellapper. 1979b. "Further Evidence on Alternative Procedures for Testing of Spatial Autocorrelation Among Regression Disturbances," pp. 113–136 in C. Bartels and R. Ketellapper (eds.), *Exploratory and Explanatory Statistical Analysis of Spatial Data.* Boston: Martinus Nijhoff.

Cliff, A. and J. Ord. 1975b. "The Comparison of Means When Samples Consist of Spatially Autocorrelated Observations," *Environment and Planning* A 7:725–734.

Doreian, P. 1981. "Estimating Linear Models With Spatially Distributed Data," pp. 359–388 in S. Leinhardt (ed.), *Sociological Methodology 1980.* San Francisco: Jossey-Bass.

Fisher, W. 1971. "Econometric Estimation With Spatial Dependence," *Regional and Urban Economics* 1:19–40.

Griffith, D. 1976. "Spatial Autocorrelation Problems: Some Preliminary Sketches of a Structural Taxonomy," *The East Lakes Geographer* 11:59–68.

Griffith, D. 1978. "A Spatially Adjusted ANOVA Model," *Geographical Analysis* 10:296–301.

Griffith, D. 1981a. "Towards a Theory of Spatial Statistics: A Rejoinder," *Geographical Analysis* 13:91–93.

Griffith, D. 1981b. "Interdependence in Space and Time: Numerical and Interpretative Considerations," pp. 258–287 in D. Griffith and R. MacKinnon (eds.), *Dynamic Spatial Models.* Alphen aan den Rijn, Netherlands: Sijthoff and Noordhoff.

Griffith, D. 1983. "The Boundary Value Problem in Spatial Statistical Analysis," *Journal of Regional Science* 23:277–287.

Griffith, D. 1985. "An Evaluation of Correction Techniques for Boundary Effects in Spatial Statistical Analysis: Contemporary Methods," *Geographical Analysis* 17:81–88.

Griffith, D. and C. Amrhein. 1983. "An Evaluation of Correction Techniques for Boundary Effects in Spatial Statistical Analysis: Traditional Methods," *Geographical Analysis.* 15: 352–360.

Haining, R. 1978a. *Specification and Estimation Problems in Models of Spatial Dependence.* Evanston, Ill.: Studies in Geography No. 24, Northwestern University Press.

Haining, R. 1978b. "Estimating Spatial-interaction Models," *Environment and Planning* A 10:305–320.

Haining, R. 1978c. "The Moving Average Model for Spatial Interaction," *Transactions,* Institute of British Geographers NS3: 202–225.

Haining, R. 1981c. "Analysing Univariate Maps," *Progress in Human Geography* 5:58–78.

Hepple, L. 1976. "A Maximum Likelihood Model for Econometric Estimation With Spatial Series," pp. 90–104 in I. Masser (ed.), *Theory and Practice in Regional Science.* London: Pion.

Hepple, L. 1978. "The Econometric Specification and Estimation of Spatio-Temporal Models," pp. 66–80 in Parkes and N. Thrift (eds.), *Time and Regional Dynamics.* New York: Halsted Press.

Hordijk, L. 1974. "Spatial Correlation in the Disturbances of a Linear Interregional Model," *Regional Science and Urban Economics* 4:117–140.

Hubert, L., R. Golledge and C. Costanzo. 1981. "Generalized Procedures for Evaluating Spatial Autocorrelation," *Geographical Analysis* 13:224–233.

Martin, R. 1974. "On Spatial Dependence, Bias, and the Use of First Spatial Differences in Regression Analysis," *Area* 6:185–194.

Miron, J. 1973. "On the Estimation of a Partial Adjustment Model With Autocorrelated Errors," *International Regional Science Review* 14:653–656.

Ord, K. 1975. "Estimation Methods for Models of Spatial Interaction," *Journal of the American Statistical Association* 70:120–126.

Streitberg, B. 1979. "Multivariate Models of Dependent Spatial Data," pp. 139–176 in P. Bartels and R. Ketellapper (eds.), *Exploratory and Explanatory Statistical Analysis of Spatial Data.* The Hague: Martinus Nijhoff.

Applications

Bannister, G. 1975. "Population Change in Southern Ontario," *Annals,* Association of American Geographers 65:177–188.

Dagbert, M. 1981. "The Simulation of Space-dependent Data in Geology," pp. 29–47 in R. Craig and M. Labovitz (eds.), *Future Trends in Geomathematics.* London: Pion.

Gatrell, A. 1979b. "The Autocorrelation Structure of Central-place Populations in Southern Germany," pp. 111–126 in N. Wrigley (editor), *Statistical Applications in the Spatial Sciences.* London: Pion.

Griffith, D. 1979. "Urban Dominance, Spatial Structure, and Spatial Dynamics: Some Theoretical Conjectures and Empirical Implications," *Economic Geography* 55:95–113.

Griffith, D. 1981c. "Modeling Urban Population Density in a Multi-Centered City," *Journal of Urban Economics* 9:298–310.

Griffith, D. 1982. "Dynamic Characteristics of Spatial Economic Systems," *Economic Geography* 58:177–196.

Griffith, D. 1983. "Phasing-out of the Sugar Industry in Puerto Rico," pp. 196–228 in D. Griffith and A. Lea (eds.), *Evolving Geographical Structures,* The Hague: Martinus Nijhoff.

Griffith, D. and K. Jones. 1980. "Explorations Into the Relationship Between Spatial Structure and Spatial Interaction," *Environment and Planning* A 12:187–201.

Haining, R. 1978d. "A Spatial Model for High Plains Agriculture," *Annals,* Association of American Geographers 68: 493–504.

Haining, R. 1981a. "Spatial Interdependencies in Population Distributions: A Study in Univariate Map Analysis. 1. Rural Population Densities," *Environment and Planning* A 13:65–84.

Haining, R. 1981b. "Spatial Interdependencies in Population Distributions: A Study in Univariate Map Analysis. 2. Urban Population Distributions," *Environment and Planning* A 13:85–96.

Haining, R. 1983. "Modelling Intra-urban Price Competition: An Example of Gasoline Pricing," *Journal of Regional Science* 23:517–528.

McCamley, F. 1972. "Testing for Spatially Autocorrelated Disturbances With Application to Relationships Estimated Using Missouri County Data," *Regional Science Perspectives* 3:89–103.

Olson, J. 1975. "Autocorrelation and Visual Map Complexity," *Annals,* Association of American Geographers 65:189–204.

Sibert, J. 1975. *Spatial Autocorrelation and the Optimal Prediction of Assessed Values.* Ann Arbor, Mich.: Michigan Geographical Publication No. 14, University of Michigan.

Wartenberg, D. 1985. "Multivariate Spatial Correlation: A Method for Exploratory Geographical Analysis," *Geographical Analysis* 17:263–283.

Other References

Berry, B. 1964. "Approaches to Regional Analysis: A Synthesis." *Annals,* Association of American Geographers 54:2–11.

Brassel, K. and D. Reif. 1979. "A Procedure to Generate Thiessen Polygons," *Geographical Analysis* 11:289–303.

Buffalo [NY] Police Department. 1982. *Annual Report 1981.* Buffalo, NY: City of Buffalo.

Curry, L. 1972. "A Spatial Analysis of Gravity Flows," *Regional Studies* 6: 131–147.

Fotheringham, A. 1981. "Spatial Structure and Distance-decay Parameters," *Annals,* Association of American Geographers 17:425–436.

Fotheringham, A. 1984. "Spatial Flows and Spatial Pattern," *Environment and Planning* A 16:529–543.

Gould, P. 1970. "Is 'Statistix Inferens' the Geographical Name for a Wild Goose?" *Economic Geography* 46 (Supplement): 439–448.

Gould, P. 1975. "Acquiring Spatial Information," *Economic Geography* 51:87–99.

Koopmans, T. 1942. "Serial Correlation and Quadratic Forms in Normal Variables," *Annals of Mathematical Statistics* 13: 14–33.

Matula, D. and R. Sokal. 1980. "Properties of Gabriel Graphs Relevant to Geographic Variation Research and the Clustering of Points in the Plane," *Geographical Analysis* 12:205–222.

Pitman, E. 1937. "The 'Closest Estimates' of Statistical Parameters," *Proceedings,* Cambridge Philosophical Society 33:212–222.

Ripley, B. 1981. *Spatial Statistics.* New York: Wiley.

Shapiro, S. and M. Wilk. 1965. "An Analysis of Variance Test for Normality," *Biometrika* 52:591–611.

Sheppard, E. 1979. "Gravity Parameter Estimation," *Geographical Analysis* 11:120–132.

Smith, T. 1980. "A Central Limit Theorem for Spatial Samples," *Geographical Analysis* 12:299–324.

Steinnes, D. 1980. "Aggregation, Gerrymandering, and Spatial Econometrics," *Regional Science and Urban Economics* 10:561–569.

Stephan, F. 1934. "Sampling Errors and the Interpretation of Social Data Ordered in Time and Space," *Journal of the American Statistical Association* 29:165–166.

Summerfield, M. 1983. "Populations, Samples and Statistical Inference in Geography," *The Professional Geographer* 35:143–149.

Tobler, W. 1970. "A Computer Movie Simulating Urban Growth in the Detroit Region," *Economic Geography* 46 (Supplement): 234–240.

Appendix A

BASIC Computer Code for the Spatial Autocorrelation Simulation Experiment

```
 10 CLS
 20 DIM NUM(16)
 30 DIM NEI(16,4)
 40 FOR I=1 TO 16
 50 READ NUM(I)
 60 NEXT I
 70 FOR I=1 TO 16
 80 FOR J=1 TO 4
 90 READ NEI(I,J)
100 NEXT J
110 NEXT I
120 Z$=TIME$
130 X$=RIGHT$(Z$,2)
140 C=VAL(X$)
150 RANDOMIZE(C)
160 DIM B(16)
170 FOR A=1 TO 16
180 B(A)=INT(RND(1)*89)+10
190 NEXT A
200 FOR A=1 TO 16
210 AVE=AVE+B(A)
220 NEXT A
230 AVE=AVE/16
240 FOR A=1 TO 16
250 VAR=VAR+(B(A)–AVE)2
260 NEXT A
270 CLS
280 E=1
290 FOR T=1 TO 4
300 PRINT “_____________________________”
310 FOR A=1 TO 4
320 PRINT “|        |        |        |        |”
330 NEXT A
340 NEXT T
350 PRINT “_____________________________”
360 FOR Q=3 TO 21 STEP 6
370 FOR W=3 TO 21 STEP 5
380 LOCATE W,Q
390 PRINT B(E)
400 LOCATE W+2,Q+1
410 PRINT CHR$(E+64)
420 E=E+1
430 NEXT W
440 NEXT O
450 LOCATE 11,30:PRINT”
```

```
460 LOCATE 12,35:PRINT"                                                  "
470 IF U=3 THEN RETURN
480 GOSUB 980
490 ZX=23
500 LOCATE 1,30
510 PRINT "DO YOU WANT TO CHANGE ANY?";
520 INPUT R$
530 IF R$="NO" THEN 1160
540 LOCATE 2,30
550 IF R$<>"YES" THEN PRINT "YES OR NO":PRINT CHR$(7):GOTO 500
560 LOCATE 1,30
570 PRINT "WHICH TWO (BOX LETTER)                              "
580 LOCATE 2,30
590 PRINT "                    "
600 LOCATE 1,55
610 INPUT D$
620 GOSUB 1170
630 LOCATE 1,60
640 INPUT S$
650 GOSUB 1200
660 IF S>16 or D>16 THEN LOCATE 3,30:PRINT CHR$(7);"A LETTER BEFORE Q
    PLEASE": GOTO 560
670 IF D<1 OR S<1 THEN LOCATE 4,30:PRINT CHR$(7);"A LETTER AFTER A
    PLEASE": GOTO 560
680 LOCATE 3,30
690 PRINT "                    "
700 LOCATE 4,30
710 PRINT "                    "
720 Y=B(D)
730 II=B(S)
740 B(D)=B(S)
750 B(S)=Y
760 U=3
770 GOSUB 270
780 ZY=1
790 GR2=GR
800 LOCATE 11,30:PRINT"IF YOU DO NOT SWITCH"
810 LOCATE 12,35:PRINT"THE CORRELATION WILL BE ";GR2
820 GOSUB 980
830 LOCATE 1,30
840 PRINT "DO YOU WANT TO KEEP THESE CHANGES?"
850 LOCATE 1,66
860 INPUT O$
870 IF O$="YES" THEN 1090
880 IF O$="NO" THEN 920
890 LOCATE 2,30
900 PRINT CHR$(7);"YES OR NO"
910 GOTO 830
920 LOCATE 2,30
930 ZY=O
940 PRINT "                    "
950 B(D)=Y
960 B(S)=II
```

```
970 GOSUB 270
980 DIF=O
990 FOR I=1 TO 16
1000 FOR J=1 TO NUM(I)
1010 DIF=DIF+(B(I)-B(NEI(I,J)))^2
1020 NEXT J
1030 NEXT I
1040 LOCATE 10,30
1050 GR=1-(((5*DIF)/(32*VAR))^.9145063)
1060 PRINT "THE CORRELATION EQUALS ";GR
1070 If ZY=1 THEN RETURN
1080 IF ZX<>23 THEN RETURN
1090 LOCATE 1,30
1100 PRINT "WOULD YOU LIKE TO MAKE SOME MORE CHANGES?
1110 LOCATE 1,72
1120 INPUT L$
1130 IF L$="NO" THEN 1160
1140 IF L$<>"YES" THEN PRINT CHR$(7):GOTO 1090
1150 GOTO 560
1160 END
1170 IF D$="" THEN RETURN
1180 D=ASC(D$)-64
1190 RETURN
1200 IF S$="" THEN RETURN
1210 S=ASC(S$)-64
1220 RETURN
1230 DATA 2,3,3,2,3,4,4,3,3,4,4,3,2,3,3,2
1240 DATA 2,5,0,0,1,3,6,0,2,4,7,0,3,8,0,0,1,6,9,0,2,5,7,10,3,6,8,11,4,7,12,0
1250 DATA 5,10,13,0,6,9,11,14,7,10,12,15,8,11,16,0,9,14,0,0
1260 DATA 10,13,15,0,11,14,16,0,12,15,0,0
```

Appendix B

Derivation of Maximum Likelihood Estimates of ρ, μ, and σ for a Spatially Autoregressive Model

$$\ell n(L) = \ell n[\prod_{i=1}^{i=n} (1 - \rho \lambda_i)] + n(\sigma_u \sqrt{2\pi})^{-n}$$
$$- \sum_{i=1}^{i=n} [(x_i - \rho \sum_{j=1}^{j=n} w_{ij} x_j) - \mu_u]^2/(2 \sigma_u^2)$$
$$= \sum_{i=1}^{i=n} \ell n(1 - \rho \lambda_i) - n \, \ell n(\sigma_u) - n \, \ell n \sqrt{2\pi}$$
$$- \sum_{i=1}^{i=n} [(x_i - \rho \sum_{j=1}^{j=n} w_{ij} x_j) - \mu_u]^2/(2 \sigma_u^2) . \quad \text{(B1)}$$

Estimating μ_u,

$$\partial \ell n(L)/\partial \mu_u = 0 - 0 - 0 - \sum_{i=1}^{i=n} 2[(x_i - \rho \sum_{j=1}^{j=n} w_{ij} x_j) - \mu_u](-1)/(2 \sigma_u^2) = 0$$

$$\sum_{i=1}^{i=n} [(x_i - \rho \sum_{j=1}^{j=n} w_{ij} x_j) - \mu_u] = 0$$

$$\sum_{i=1}^{i=n} (x_i - \rho \sum_{j=1}^{j=n} w_{ij} x_j) - \sum_{i=1}^{i=n} \mu_u = 0$$

$$n \, \mu_y = \sum_{i=1}^{i=n} (x_i - \rho \sum_{j=1}^{j=n} w_{ij} x_j)$$

$$\therefore \hat{\mu}_u = \sum_{i=1}^{i=n} (x_i - \rho \sum_{j=1}^{j=n} w_{ij} x_j)/n .$$

Next, estimating σ_u,

$$\partial \ell n(L)/\partial \sigma_u = 0 - n/\sigma_u - 0 + \sum_{i=1}^{i=n} [(x_i - \rho \sum_{j=1}^{j=n} w_{ij} x_j) - \mu_u]^2/(\sigma_u^3) = 0$$

$$- n/\sigma_u + \sum_{i=1}^{i=n} [(x_i - \rho \sum_{j=1}^{j=n} w_{ij} x_j) - \mu_u]^2/(\sigma_u^3) = 0$$

$$- n \, \sigma_u^2 + \sum_{i=1}^{i=n} [(x_i - \rho \sum_{j=1}^{j=n} w_{ij} x_j) - \mu_u]^2 = 0$$

$$\therefore \hat{\sigma}_u^2 = \sum_{i=1}^{i=n} [(x_i - \rho \sum_{j=1}^{j=n} w_{ij} x_j) - \mu_u]^2/n.$$

Finally, estimating ρ,

$$\partial \ell n(L)/\partial \rho = \sum_{i=1}^{i=n} - \lambda_i/(1 - \rho \lambda_i) - 0 - 0$$
$$- \sum_{i=1}^{i=n} 2[(x_i - \rho \sum_{j=1}^{j=n} w_{ij} x_j) - \mu_u](- \sum_{j=1}^{j=n} w_{ij} x_j)/(2 \sigma_u^2) . \quad \text{(B2)}$$

Unfortunately this partial derivative does not reduce to a neat, linear form, as did those for μ_u and σ_u. Consequently, nothing is gained by studying this derivative. Rather, the original log-likelihood function needs to be subjected to nonlinear optimization techniques.

Returning to equation (B1), if σ_u^2 is substituted into this equation, then it becomes

$$\ell n(L) = \ell n[\prod_{i=1}^{i=n} (1 = \rho \lambda_i)]$$
$$- n\,\ell n\{ \sum_{i=1}^{i=n} [(x_i - \rho \sum_{j=1}^{j=n} w_{ij}\, x_j) - \mu_u]^2/n\}^{1/2} - n\,\ell n \sqrt{2\pi}$$
$$- \sum_{i=1}^{i=n} [(x_i - \rho \sum_{j=1}^{j=n} w_{ij}\, x_j) - \mu_u]^2/\{2 \sum_{i=1}^{i=n} [(x_i - \rho \sum_{j=1}^{j=n} w_{ij}\, x_j) - \mu_u]^2/n\}$$
$$= \ell n\, [\prod_{i=1}^{i=n} (1 - \rho \lambda_i)] - n\,\ell n\{n^{-1} \sum_{i=1}^{i=n} [(x_i - \rho \sum_{j=1}^{j=n} w_{ij}\, x_j) - \mu_u]^2\}/2$$
$$- n\,\ell n \sqrt{2\pi} - n/2\,.$$

Next, substituting μ_u into this equation, and factoring out (1/2) yields

$$\ell n(L) = (-1/2)\left[-2\,\ell n\, [\prod_{i=1}^{i=n} (1 - \rho \lambda_i)] - n\,\ell n(n) +\right.$$
$$n\,\ell n\{ \sum_{i=1}^{i=n} [(x_i - \rho \sum_{j=1}^{j=n} w_{ij}\, x_j) - \sum_{i=1}^{i=n} (x_i - \rho \sum_{j=1}^{j=n} w_{ij}\, x_j)/n]^2\}$$
$$\left. + 2n\,\ell n \sqrt{2\pi} + n \right]$$
$$= (-1/2)[(2\,n\,\ell n \sqrt{2\pi} + n - n\,\ell n(n)] +$$
$$(-1/2)\,[\ell n\, \{\prod_{i=1}^{i=n} (1 - \rho \lambda_i)\}^{-2}$$
$$+ n\,\ell n\{ \sum_{i=1}^{i=n} [(x_i - \sum_{i=1}^{i=n} x_i/n) - \rho \sum_{j=1}^{j=n} w_{ij}\, x_j +$$
$$\rho \sum_{i=1}^{i=n} \sum_{j=1}^{j=n} w_{ij}\, x_j/n]^2\}\,.$$

The first term of this equation consists entirely of constants, and hence will not affect the minimization of $\ell n(L)$ except as a scaling factor, say K. The second term will govern the minimization. Factoring an n out of this term, and taking the anti-logarithm of the entire equation yields

$$L = K\{[\prod_{i=1}^{i=n} (1 - \rho \lambda_i)]^{-2/n} \sum_{i=1}^{i=n} [(x_i - \bar{x}) -$$
$$\rho(\sum_{j=1}^{j=n} w_{ij}\, x_j - \sum_{i=1}^{i=n} \sum_{j=1}^{j=n} w_{ij}\, x_j/n)]^2\}^{-n/2}\,. \qquad \text{(B3)}$$

The expression contained within the braces of this version of the likelihood function is what Meade (1967) maintains should be minimized, in order for L to be maximized.

Maximization of equation (B3) is facilitated by rewriting the squared term in its quadratic form, implying that the expression to be minimized should be

$$[\prod_{i=1}^{i=n} (1 - \rho \lambda_i)]^{-2/n} [\sum_{i=1}^{i=n} (x_i - \bar{x})^2 - 2\rho \sum_{i=1}^{i=n} (x_i - \bar{x})(\sum_{j=1}^{j=n} w_{ij} x_j$$

$$- \sum_{i=1}^{i=n} \sum_{j=1}^{j=n} w_{ij} x_j/n) + \rho^2 \sum_{i=1}^{i=n} (\sum_{j=1}^{j=n} w_{ij} x_j - \sum_{i=1}^{i=n} \sum_{j=1}^{j=n} w_{ij} x_j/n)^2]. \qquad \text{(B4)}$$

In this form the λ_is and summation terms can be calculated directly from the data, and then substituted into equation (B4) so that only the ρ value is unknown.

Appendix C

A Derivation of the Expected Value of the Geary Ratio Under the Randomization Assumption

Because the configuration of areal units is fixed, and the same values are being allocated to the areal units during each permutation,

$$E\{(n-1)\sum_{i=1}^{i=n}\sum_{j=1}^{j=n} c_{ij}(x_i - x_j)^2/[2\sum_{i=1}^{i=n}\sum_{j=1}^{j=n} c_{ij}(\sum_{i=1}^{i=n} x_i - \bar{x})^2]\}$$

$$= (n-1)\sum_{i=1}^{i=n}\sum_{j=1}^{j=n} c_{ij}\, E[(x_i - x_j)^2]/[2\sum_{i=1}^{i=n}\sum_{j=1}^{j=n} c_{ij}(\sum_{i=1}^{i=n} x_i - \bar{x})^2]$$

$$= (n-1)\, E(x_i - \bar{x} + \bar{x} - x_j)^2]\sum_{i=1}^{i=n}\sum_{j=1}^{j=n} c_{ij}/[2\sum_{i=1}^{i=n}\sum_{j=1}^{j=n} c_{ij}\sum_{i=1}^{i=n}(x_i - \bar{x})^2]$$

$$= (n-1)\{E[(x_i - \bar{x})^2] - 2\,E[(x_i - \bar{x})(x_j - \bar{x})] +$$

$$E[(x_j - \bar{x})^2]\}/[2\sum_{i=1}^{i=n}(x_i - \bar{x})^2]$$

$$= (n-1)\{[\sum_{i=1}^{i=n}(x_i - \bar{x})^2/n]/[\sum_{i=1}^{i=n}(x_i - \bar{x})^2]$$

$$- 2(-1/n-1)[\sum_{i=1}^{i=n}(x_i - \bar{x})^2/n]/[\sum_{i=1}^{i=n}(x_i - \bar{x})^2] +$$

$$[\sum_{i=1}^{i=n}(x_i - \bar{x})^2/n]\}/[2\sum_{i=1}^{i=n}(x_i - \bar{x})^2$$

$$= (n-1)\{1/n + 2/[n(n-1)] + 1/n\}/2$$

$$= 1\,.$$

$E[(x_i - \bar{x})^2] = \sum_{i=1}^{i=n}(x_i - \bar{x})^2/n$ because this second term is constant over all permutations, since the x_is are fixed. And, as Cliff and Ord (1981a) show,

$$E[(x_i - \bar{x})(x_j - \bar{x})] = \sum_{i=1}^{i=n}(x_i - \bar{x})[\sum_{j=1}^{j=n}(x_j - \bar{x}) - (x_i - \bar{x})]/(n-1)\,,$$

since once x_i is selected there can be only $(n-1)$ possible x_js to consider. But, $\sum_{i=1}^{i=n}(x_j - \bar{x}) = 0$, and thus

$$E[(x_i - \bar{x})(x_j - \bar{x})] = -\sum_{i=1}^{i=n}(x_i - \bar{x})^2/(n-1)\,.$$

Appendix D

A Derivation of the Expected Value of the Geary Ratio Under the Normality Assumption

Utilizing the Pitman-Koopmans theorem (Pitman 1937; Koopmans 1942),

$$E\{(n-1)\sum_{i=1}^{i=n}\sum_{j=1}^{j=n} c_{ij}(x_i - x_j)^2/[2\sum_{i=1}^{i=n}\sum_{j=1}^{j=n} c_{ij}\sum_{i=1}^{i=n}(x_i - \bar{x})^2]\}$$

$$= (n-1)\sum_{i=1}^{i=n}\sum_{j=1}^{j=n} c_{ij}\, E[(x_i - x_j)^2]/[2\sum_{i=1}^{i=n}\sum_{j=1}^{j=n} c_{ij}\, E[(x_i - \bar{x})^2]$$

$$= (n-1)\, E[(x_i - \bar{x} + \bar{x} - x_j)^2]\sum_{i=1}^{i=n}\sum_{j=1}^{j=n} c_{ij}/[2\sum_{i=1}^{i=n}\sum_{j=1}^{j=n} c_{ij}(n-1)\,\sigma^2]$$

$$= E[(x_i - \bar{x})^2 - 2(x_i - \bar{x})(x_j - \bar{x}) + (x_j - \bar{x})^2]/(2\,\sigma^2)$$

$$= [(n-1)\sigma^2/n - 2(-\sigma^2 n) + (n-1)\sigma^2/n]/(2\,\sigma^2)$$

$$= (n-1)/(2n) + 2/(2n) + (n-1)/(2n) = 1\,.$$

Appendix E

A Derivation of the Expected Value of the Moran Coefficient Under the Randomization and Normality Assumptions

If the randomization assumption prevails, then

$$E\{n\sum_{i=1}^{i=n}\sum_{j=1}^{j=n} c_{ij}(x_i - \bar{x})(x_j - \bar{x})/[\sum_{i=1}^{i=n}\sum_{j=1}^{j=n} c_{ij}\sum_{i=1}^{i=n}(x_i - \bar{x})^2]\}$$

$$= n\, E[(x_i - \bar{x})(x_j - \bar{x})]\sum_{i=1}^{i=n}\sum_{j=1}^{j=n} c_{ij}/[\sum_{i=1}^{i=n}\sum_{j=1}^{j=n} c_{ij}\sum_{i=1}^{i=n}(x_i - \bar{x})^2]$$

$$= n[-1/(n-1)][\sum_{i=1}^{i=n}(x_i - \bar{x})^2/n]/[\sum_{i=1}^{i=n}(x_i - \bar{x})^2]$$

$$= -\,1/(n-1)\,.$$

On the other hand, if the normality assumption prevails, then

$$E\{n\sum_{i=1}^{i=n}\sum_{j=1}^{j=n} c_{ij}(x_i - \bar{x})(x_j - \bar{x})/[\sum_{i=1}^{i=n}\sum_{j=1}^{j=n} c_{ij}\sum_{i=1}^{i=n}(x_i - \bar{x})^2]\}$$

$$= n\sum_{i=1}^{i=n}\sum_{j=1}^{j=n} c_{ij}\, E[(x_i - \bar{x})(x_j - \bar{x})]/[\sum_{i=1}^{i=n}\sum_{j=1}^{j=n} c_{ij}(n-1)\,\sigma^2]$$

$$= n\,(-\sigma^2/n)\sum_{i=1}^{i=n}\sum_{j=1}^{j=n} c_{ij}/[\sum_{i=1}^{i=n}\sum_{j=1}^{j=n} c_{ij}(n-1)\,\sigma^2] = -\,1/(n-1)\,.$$